AN INTRODUCTION TO
NUMERICAL MATHEMATICS

Academic Press
Textbooks in Mathematics

Consulting Editor: **Ralph P. Boas, Jr.,** Northwestern University

---

HOWARD G. TUCKER. An Introduction to Probability and Mathematical Statistics

EDUARD L. STIEFEL. An Introduction to Numerical Mathematics

PESI MASANI. Elementary Calculus

WILLIAM PERVIN. Foundations of General Topology

JAMES SINGER. Fundamentals of Numerical Analysis

AN INTRODUCTION TO

NUMERICAL
MATHEMATICS

Eduard L. Stiefel

Department of Mathematics
Eidgenoessische Technische Hochschule
Zürich, Switzerland

Translated by

Werner C. Rheinboldt

Computer Science Center
University of Maryland
College Park, Maryland

and

Cornelie J. Rheinboldt

NEW YORK **ACADEMIC PRESS** LONDON

ACADEMIC PRESS INC.
111 Fifth Avenue, New York 3, New York

United Kingdom Edition published by
ACADEMIC PRESS INC. (LONDON) LTD.
Berkeley Square House, London W.1

LIBRARY OF CONGRESS CATALOG CARD NUMBER: **63-21407**

PRINTED IN THE UNITED STATES OF AMERICA

AN INTRODUCTION TO
NUMERICAL MATHEMATICS

THIS VOLUME WAS ORIGINALLY PUBLISHED
BY B. G. TEUBNER, VERLAGSGESELLSCHAFT, STUTTGART
AS EINFUHRUNG IN DIE NUMERISCHE MATHEMATIK.
AUTHOR'S REVISIONS OF THE FIRST EDITION HAVE
BEEN INCLUDED IN THIS TRANSLATION.

Foreword to German Edition

This book has grown out of courses which the author gave at the Swiss Federal Institute of Technology (ETH) and elsewhere. As was the case with the lectures themselves, this work is directed not only to mathematicians and physicists, but to a wider circle of interested persons as well—and particularly to engineers. For this reason an elementary form of presentation was chosen by not assuming a knowledge of linear algebra. Rather, the necessary theoretical results are derived as a by-product of the discussion of numerical methods. The formalism of matrix-calculus was not used; besides, it would not have lent itself well to the treatment of linear programming.

Up to the beginning of chapter 6, the mathematical background generally taught in the first years of a college science curriculum is sufficient for an understanding of the material. Later on it becomes desirable that the reader has mastered the fundamentals of the theory of differential equations. In this connection the book by L. Collatz "Differential Equations for Engineers" from the same publisher is especially recommended.

This book is based completely on the algorithmic standpoint. Interest has therefore been directed toward certain computational procedures selected in such a way that they cover first of all a wider range of applications, and are furthermore suitable for automatic computation. As an example, the first three parts have been devoted to one single algorithmic method, namely the exchange-method, which has also been used in theoretical connection by Steinitz for the foundation of linear algebra. This method is a shining illustration of the fact that there is only one single field of mathematics and not two separate ones—pure and applied. The proof-techniques derived by the theoreticians can be expanded just as often into practically usable algorithms with little effort, but with the necessary insight. The reverse point, namely the fact that the invention of a computational method can lead to a theoretical insight, should be evident in a number of places throughout this book. The mentioned algorithmic standpoint is also the recurring theme of the newer computation centers. They try to meet the recurring needs of their customers with computer programs which were set up once and for all, rather than first developing ad hoc the optimal numerical method for every problem presented to them.

The book's elementary character, however, has not precluded presentation of some newer results. In this connection we need only point to the Chebyshev approximation, the digression into game theory, the rhombus algorithm (quotient difference algorithm), and to the numerical integration by repeated halving of the step. Over and against this, other things had to retreat more into the background; for example, the interpolation theory was discussed in a very condensed form only.

I should like to thank my colleagues Professor Dr. P. Henrici and Professor Dr. H. Rutishauser, for their stimulating discussions, and also Mr. H. Bührer for his work on the manuscript and the index. Finally, I want to thank the publishers for their careful production of the book, and considerate printing.

E. STIEFEL.

Zürich, June 1963.

Foreword to the English Edition

The first German edition of this book by E. L. Stiefel has met with such widespread success that, within less than one year, a second edition was indicated. The preparation of this second German edition—which appeared earlier ths year—parallels the translation of the work into English, and as a result, all changes and revisions contained in the second edition have also been incorporated in the translation.

Because of the very short time period involved since the first publication of the book, the author made no major changes. Occasionally, the text was simplified, expanded, or slightly modified in style. The only new parts introduced into the second edition—and also incorporated into the English translation—are the short presentation of the Everett interpolation formula, the Adams method with interpolation, and the over-relaxation method. The coordinate nomogram of Appendix II was changed to an alignment-nomogram.

With the approval of the author, the translators supplemented the English edition with a bibliography of texts available in the English language.

W. C. RHEINBOLDT
C. J. RHEINBOLDT

Contents

ix

I Linear Algebra

The most important problem of numerical linear algebra is the development of algorithms—that is, of arithmetical procedures—for the solution of systems of linear equations with many unknowns. Essentially there are two reasons why this is such a basic problem. First of all, today nearly every linear problem in applied mathematics—as for example, a boundary-value problem for a linear ordinary or partial differential equation—is reduced by appropriate techniques to a system of linear equations, especially when electronic computers are to be employed for its solution. Secondly, nonlinear problems can in most cases be solved only by approximating them by linear problems.

I.I Linear Functions, Exchange

For simplicity's sake, we shall begin with a linear function y of two independent variables, x, x';

$$y = ax + bx' + k. \tag{1}$$

Here, a, b are fixed given coefficients and the constant k is also a given number. In a three-dimensional coordinate system with the axes x, x', y the linear function is represented by a plane. When k equals zero, the function

$$y = ax + bx' \tag{2}$$

is called a *homogeneous linear function*, or simply a *linear form*. The above-mentioned plane then passes through the origin of the coordinate system. Now let

$$y' = cx + dx' \tag{3}$$

be a second linear form in the same variables x and x'. Assuming that $a \neq 0$, we can calculate x explicitly from the first form, insert it into the second and so obtain

$$x = \frac{1}{a}y - \frac{b}{a}x'$$

$$y' = \frac{c}{a}y + \left(d - \frac{bc}{a}\right)x'. \tag{4}$$

1

The result then is two new linear forms which are characterized by the fact that the formerly independent variable x has now become a function, while y has become an independent variable. In short, we can say that x and y have been exchanged and we call the entire operation an *exchange-step* or, in brief, *EX-step*. This EX-operation forms the basis of nearly all algorithms of linear algebra and in most cases these are constructed of many EX-steps. Our problem now will be to arrange such constructions in the form of clear computing schemes. For example, the above EX-step will henceforth be represented by the following scheme:

$$
\begin{array}{c}
\quad x \quad x' \\
\begin{array}{c} y = \\ y' = \end{array}
\boxed{\begin{array}{cc} a & b \\ c & d \end{array}}
\end{array}
\qquad
\begin{array}{c}
\qquad\quad y' \qquad\qquad x' \\
\begin{array}{c} x = \\[2em] y' = \end{array}
\boxed{\begin{array}{cc} \dfrac{1}{a} & -\dfrac{b}{a} \\[1em] \dfrac{c}{a} & d - \dfrac{bc}{a} \end{array}}
\end{array}
\tag{5}
$$

The column x and the row y of the variables being exchanged intersect in the element a; this element shall be called the *pivot element* of the step. The element d diagonal to a in the rectangle, is subjected to the most complicated transformation; the correction term bc/a is obtained as follows: The elements forming the co-diagonal of the rectangle are to be multiplied and the result is to be divided by the pivot element. This rule will be called the *rectangle rule*.

Let us now consider several linear forms in several variables; for example, in our schematic notation:

$$
\begin{array}{c}
\quad x_1 \quad\; x_2 \quad\; x_3 \quad\; x_4 \\
\begin{array}{c} y_1 = \\ y_2 = \\ y_3 = \end{array}
\boxed{\begin{array}{cccc}
a_{11} & a_{12} & a_{13} & a_{14} \\
a_{21} & a_{22} & a_{23} & a_{24} \\
a_{31} & a_{32} & a_{33} & a_{34}
\end{array}}
\end{array}
\tag{6}
$$

Written in full, these equations have, for example, the form

$$y_2 = a_{21}x_1 + a_{22}x_2 + a_{23}x_3 + a_{24}x_4$$

or, more generally,

$$y_i = \sum_{k=1}^{4} a_{ik}x_k, \qquad i = 1, 2, 3. \tag{7}$$

The block (6) of coefficients is called the matrix of the given linear forms. We will now exchange any one of the independent variables with any given dependent variable. For example, let us proceed with the exchange of x_3 and y_2. The pivot column x_3 and the pivot row y_2 intersect in the pivot element a_{23}; the exchange is possible if this element is different from zero. The variable x_3 now plays the role of x in our elementary EX-step (5), whereas y_2 assumes the role of y. All other variables x_1, x_2, x_4 and y_1, y_3 not participating in the exchange, behave like x' and y' in (5). Application of these remarks results directly in the scheme following the exchange

$$
\begin{array}{c|cccc}
 & x_1 & x_2 & y_2 & x_4 \\
\hline
y_1 = & \alpha_{11} & \alpha_{12} & \dfrac{a_{13}}{a_{23}} & \alpha_{14} \\[2ex]
x_3 = & -\dfrac{a_{21}}{a_{23}} & -\dfrac{a_{22}}{a_{23}} & \dfrac{1}{a_{23}} & -\dfrac{a_{24}}{a_{23}} \\[2ex]
y_3 = & \alpha_{31} & \alpha_{32} & \dfrac{a_{33}}{a_{23}} & \alpha_{34}
\end{array}
\tag{8}
$$

where

$$\alpha_{11} = a_{11} - \frac{a_{21}a_{13}}{a_{23}}, \qquad \alpha_{12} = a_{12} - \frac{a_{22}a_{13}}{a_{23}}, \qquad \alpha_{14} = a_{14} - \frac{a_{24}a_{13}}{a_{23}},$$

$$\alpha_{31} = a_{31} - \frac{a_{21}a_{33}}{a_{23}}, \qquad \alpha_{32} = a_{32} - \frac{a_{22}a_{33}}{a_{23}}, \qquad \alpha_{34} = a_{34} - \frac{a_{24}a_{33}}{a_{23}},$$

$$\tag{8a}$$

which can also be checked by explicit calculations.

We can summarize the rules for such an EX-step as follows:

1. *The pivot element is replaced by its reciprocal value.*
2. *The other elements of the pivot column are to be divided by the pivot element.*
3. *The other elements of the pivot row are to be divided by the pivot element and given the opposite sign.*
4. *An element in the remaining part of the matrix is transformed by applying the rectangle rule to that rectangle of 4 matrix elements which contains the element itself in one corner and the pivot element in the corner diagonal to that.*

Exchanging is a reversible process; in fact, one obtains scheme (6) from (8) by applying an EX-step with the pivot $(1/a_{23})$. Hence, the relations (6) between the variables x and y are a consequence of the relations (8). In other words, if there are 7 given values of the variables

$$x_1, x_2, x_3, x_4; \qquad y_1, y_2, y_3$$

which satisfy the relations (8), then they also satisfy (6).

Numerical example of an exchange step:

<table>
<tr><td></td><td>x_1</td><td>$\underline{x_2}$</td><td>x_3</td><td></td><td></td><td>x_1</td><td>y_3</td><td>x_3</td></tr>
<tr><td>$y_1 =$</td><td>3</td><td>$\underline{5}$</td><td>1</td><td>$\longrightarrow$</td><td>$y_1 =$</td><td>0.5</td><td>2.5</td><td>-4</td></tr>
<tr><td>$y_2 =$</td><td>2</td><td>4</td><td>5</td><td></td><td>$y_2 =$</td><td>0</td><td>2</td><td>1</td></tr>
<tr><td>$\underline{y_3} =$</td><td>$\underline{1}$</td><td>$\underset{=}{2}$</td><td>$\underline{2}$</td><td></td><td>$x_2 =$</td><td>-0.5</td><td>0.5</td><td>-1</td></tr>
</table>

$$-0.5 \qquad\qquad -1 \tag{9}$$

To facilitate the calculation, the new pivot row was written beneath the old matrix (but with the element in the pivot column omitted). We call this row the basement row. It can be easily verified that, instead of the above Rule 4, we may use the simpler rule:

4′. *An element in the remaining part of the matrix is transformed by adding to it the product formed with the element vertically beneath it in the basement row as first factor, and that element in the pivot column which is in the same row as the given element as second factor.*

Calculating according to this rule, we observe that the evaluation of each new element involves exactly one multiplication (or division). If we regard multiplications and divisions as the essential arithmetic operations whose number determines the time consumption of the calculation, we note that:

An EX-step for m functions with n variables requires $m \cdot n$ essential operations.

1.11   Control Sums

Let us assume for a moment that the sum of the elements in every row of a given matrix equals 1. This means: Given the value 1 for all independent variables x, all dependent variables y will have the value 1 as well. This situation also holds true after an exchange-step, and hence, following such a step, the row sum again is equal to 1. For control purposes it is possible to enforce the value 1 for the row sums by introducing an additional variable σ next to the x in the linear forms with its coefficient selected in such a way that indeed all row sums equal 1.

In example (9) this looks as follows:

	x_1	x_2	x_3	σ		x_1	y_3	x_3	σ
$y_1 =$	3	5	1	-8	$y_1 =$	0.5	2.5	-4	2
$y_2 =$	2	4	5	-10	$y_2 =$	0	2	1	-2
$y_3 =$	1	2	2	-4	$x_2 =$	-0.5	0.5	-1	2
	-0.5	-1	2						

$$(10)$$

In an EX-step the σ column is treated in the same way as any x-column, namely, according to Rules 1 through 4'. The control of the calculation consists of the fact that in the new scheme all row sums must again be equal to 1.

A more extensive numerical example of the EX-method is given in Appendix I, Example 1.

1.12 Transposed Notation

Linear forms can also be obtained from a given matrix (6) by reading this matrix in the direction of the columns instead of the rows. We will establish the following rule for this ordering:

	$v_1 =$	$v_2 =$	$v_3 =$	$v_4 =$
$-u_1$	a_{11}	a_{12}	a_{13}	a_{14}
$-u_2$	a_{21}	a_{22}	a_{23}	a_{24}
$-u_3$	a_{31}	a_{32}	a_{33}	a_{34}

$$(11)$$

In other words, the independent variables u_1, u_2, u_3 on the left border are to be taken with the negative sign. Written out in full, for example,

$$v_2 = -a_{12}u_1 - a_{22}u_2 - a_{32}u_3$$

and, more generally,

$$v_i = - \sum_{k=1}^{3} a_{ki}u_k, \qquad i = 1, 2, 3, 4.$$

$$(12)$$

These forms v_i are called the transposed forms of the original forms (7). Here, as in (6), we will perform the exchange-step with the pivot a_{23}; that is, we will make v_3 an independent variable and u_2 a dependent variable.

This step can be performed according to Rules 1 through 4 if the words "column" and "row" in Rules 2 and 3 are interchanged. (The rectangle rule is symmetric and therefore invariant with respect to interchanges of rows and columns.) In this way, we obtain the following scheme:

	$v_1 =$	$v_2 =$	$-u_2 =$	$v_4 =$
$-u_1$	α_{11}	α_{12}	$-\dfrac{a_{13}}{a_{23}}$	α_{14}
v_3	$\dfrac{a_{21}}{a_{23}}$	$\dfrac{a_{22}}{a_{23}}$	$\dfrac{1}{a_{23}}$	$\dfrac{a_{24}}{a_{23}}$
$-u_3$	α_{31}	α_{32}	$-\dfrac{a_{33}}{a_{23}}$	α_{34}

where the elements α are again given by (8a). If we now reverse the signs of the variables involved in the exchange, we obviously get

	$v_1 =$	$v_2 =$	$u_2 =$	$v_4 =$
$-u_1$	α_{11}	α_{12}	$\dfrac{a_{13}}{a_{23}}$	α_{14}
$-v_3$	$-\dfrac{a_{21}}{a_{23}}$	$-\dfrac{a_{22}}{a_{23}}$	$\dfrac{1}{a_{23}}$	$-\dfrac{a_{24}}{a_{23}}$
$-u_3$	α_{31}	α_{32}	$\dfrac{a_{33}}{a_{23}}$	α_{34}

(13)

The variables on the left-hand border are again taken with the negative sign and the resulting table is exactly identical with table (8). It is, of course, also possible to verify (13) by an explicit calculation. We have obtained the important

Duality Theorem of Linear Algebra. *An exchange-step for a given set of linear forms is completely identical with that exchange-step for the transposed forms which uses the same pivot.*

This theorem constitutes the source for many additional duality theorems in linear algebra. We shall give an example of this in the section on linear programming.

1.13 Modified Exchange

For the future treatment of the symmetry properties of linear systems of equations, it will be some disadvantage that—according to Rule 3—the signs in the pivot row have to be reversed; or in other words, that the pivot row has to be treated differently from the pivot column. This can be remedied by writing (8) as follows:

$$
\begin{array}{c|cccc}
 & x_1 & x_2 & y_2 & x_4 \\
\hline
y_1 = & \alpha_{11} & \alpha_{12} & \dfrac{a_{13}}{a_{23}} & \alpha_{14} \\
-x_3 = & \dfrac{a_{21}}{a_{23}} & \dfrac{a_{22}}{a_{23}} & -\dfrac{1}{a_{23}} & \dfrac{a_{24}}{a_{23}} \\
y_3 = & \alpha_{31} & \alpha_{32} & \dfrac{a_{33}}{a_{23}} & \alpha_{34}
\end{array}
\tag{14}
$$

In a later instance we will use this minor modification for certain theoretical purposes, but never in practical computations.

1.2 Inversion

Consider n linear forms with n variables. In this case, the matrix of the linear forms is square; that is, there are as many rows as there are columns. By means of repeated EX-steps we can then attempt finally to bring all x to the left border and all y to the upper one. Example (9) will be treated in this manner. According to (9), the first exchange of x_2 and y_3 resulted in

$$
\begin{array}{c|ccc}
 & x_1 & x_2 & x_3 \\
\hline
y_1 = & 3 & 5 & 1 \\
y_2 = & 2 & 4 & 5 \\
y_3 = & 1 & 2 & 2 \\
\hline
 & -0.5 & -1 &
\end{array}
\qquad
\begin{array}{c|ccc}
 & x_1 & y_3 & x_3 \\
\hline
y_1 = & 0.5 & 2.5 & -4 \\
y_2 = & 0 & 2 & 1 \\
x_2 = & -0.5 & 0.5 & -1 \\
\hline
 & -5 & 8 &
\end{array}
\tag{15}
$$

To the left, we have the given square matrix and to the right the result of the first EX-step. In order to bring another x to the left border and another y to the upper one, the pivot must be taken from a y-row and an x-column. We therefore have 4 elements of the new matrix as possible candidates for the pivot. Selection of the element in the left-hand upper corner as pivot and execution of the corresponding exchange-step then leads to:

	y_1	y_3	$\underline{x_3}$
$x_1 =$	2	−5	8
$y_2 =$	0	2	1
$x_2 =$	−1	3	−5

         0   −2

	y_1	y_3	y_2
$x_1 =$	2	−21	8
$x_3 =$	0	−2	1
$x_2 =$	−1	13	−5

$$(16)$$

For the final exchange, we have to take as pivot the element in the intersection of the remaining y-row and x-column—we have no other choice. The third and final exchange then yields the result given on the right-hand side of (16). Written out in full, this reads as follows:

$$
\begin{aligned}
x_1 &= \ 2y_1 + 8y_2 - 21y_3 \\
x_2 &= -y_1 - 5y_2 + 13y_3 \\
x_3 &= \ y_2 - \ 2y_3
\end{aligned}
\qquad (17)
$$

Here all y have become independent, and all x dependent variables. The forms (17) are called the inverses of the given linear forms, and their matrix

$$
\begin{array}{ccc}
2 & 8 & -21 \\
-1 & -5 & 13 \\
0 & 1 & -2
\end{array}
\qquad (18)
$$

is the *inverse matrix* of the original matrix on the left-hand side of (15).

Any EX-step for an $n \times n$ matrix involves n^2 essential operations as we have seen in 1.1. For an inversion we need n EX-steps and hence we find: *The inversion of n linear forms involves n^3 essential operations.*

If, therefore, one needs half an hour to invert a 3×3 matrix, he will need some 32 hours for the inversion of a 12×12 matrix. Should he make a mistake in his calculation—having forgotten the control sums—he may well spend an entire week on this work. Appendix I, Example 1, shows the inversion of a 4×4 matrix.

1.21 Solution of Linear Equations by Inversion

As an example, let us consider the three linear equations with the three unknowns x_1, x_2, x_3:

$$\left.\begin{array}{r} 3x_1 + 5x_2 + x_3 = -4 \\ 2x_1 + 4x_2 + 5x_3 = 9 \\ x_1 + 2x_2 + 2x_3 = 3 \end{array}\right| \tag{19}$$

The 3×3 matrix formed by the coefficients of the unknowns is called the *matrix of the system of equations*; we chose this matrix to be identical with the matrix of the linear forms y_1, y_2, y_3 in (15). In other words, our equations state that y_1, y_2, y_3 have the values -4, 9, 3, respectively. Substitution of these values in the inverse forms (17) produces

$$x_1 = 1, \qquad x_2 = -2, \qquad x_3 = 3 \tag{20}$$

as the solution of the system of equations. The fact that these values indeed satisfy the given equations (19) follows readily from the reversibility of the exchange process as pointed out in 1.1. If n equations with n unknowns are to be solved, the substitution in the inverse forms requires n^2 multiplications.

Frequently the problem arises that more than one system of equations must be solved, all of which have the same coefficient-matrix, but different right-hand sides. For the solution of k such systems, we obviously must perform $(n^3 + kn^2)$ essential operations.

Particularly interesting is the special case when all right-hand sides equal zero. In our example, this would be the system of equations

$$\left.\begin{array}{r} 3x_1 + 5x_2 + x_3 = 0 \\ 2x_1 + 4x_2 + 5x_3 = 0 \\ x_1 + 2x_2 + 2x_3 = 0 \end{array}\right| \tag{21}$$

This is called the *homogeneous system* corresponding to the given system (19) (or to the given matrix). One solution of a homogeneous system can, of course, be produced directly without any calculation: if all unknowns are set equal to zero, the homogeneous equations are obviously satisfied. This solution is called the *trivial solution* or the *zero-solution*. In the theory of linear systems of equations, it is decisively important whether the given homogeneous system possesses only the trivial solution. This is the case in our example where it follows from (17) that, if all y are zero, all x will also have to be zero.

1.22 Exceptional Cases, Theoretical Consequences

As basis for the discussion in this section, we choose the inversion of the 4×4 matrix

	x_1	x_2	x_3	x_4
$y_1 =$	3	5	1	2
$y_2 =$	2	−4	3	7
$y_3 =$	4	14	−1	−3
$y_4 =$	13	7	9	20
	−3	−5		−2

	x_1	x_2	y_1	x_4
$x_3 =$	−3	−5	1	−2
$y_2 =$	−7	−19	3	1
$y_3 =$	7	19	−1	−1
$y_4 =$	−14	−38	9	2
	+7	+19	−3	

A first exchange-step has already been made here and a second one results in

	x_1	x_2	y_1	y_2
$x_3 =$	−17	−43	7	−2
$x_4 =$	7	19	−3	1
$y_3 =$	0	0	2	−1
$y_4 =$	0	0	3	2

$$(22)$$

A continuation of the inversion is now impossible. All candidates for the next pivot—which must stand in a y-row and an x-column—are equal to zero. However, an exchange-step can be carried out only when the pivot element differs from zero. A number of conclusions can be drawn from this situation. First of all, the remaining y-rows yield the relations

$$y_3 = 2y_1 - y_2, \qquad y_4 = 3y_1 + 2y_2$$

Hence, the given linear forms are mutually dependent. For example, the third one is twice the first minus the second, as can easily be verified from the given matrix. The EX-algorithm therefore permits such dependencies to be determined. Secondly, the x-rows give

$$x_3 = -17x_1 - 43x_2 + 7y_1 - 2y_2$$
$$x_4 = 7x_1 + 19x_2 - 3y_1 + y_2$$

which can be regarded as at least a partial inversion.

Let us now solve the homogeneous equations corresponding to the given matrix

$$3x_1 + 5x_2 + x_3 + 2x_4 = 0$$

$$2x_1 - 4x_2 + 3x_3 + 7x_4 = 0$$

$$4x_1 + 14x_2 - x_3 - 3x_4 = 0$$

$$13x_1 + 7x_2 + 9x_3 + 20x_4 = 0$$

Setting y_1 through y_4 equal to zero in (22) we obtain

$$x_3 = -17x_1 - 43x_2$$
$$x_4 = 7x_1 + 19x_2. \tag{23}$$

If we arbitrarily select x_1 and x_2 and calculate x_3, x_4, the reversibility of the EX-process guarantees that these 4 values of the unknowns satisfy the homogeneous equations. The homogeneous system of equations therefore possesses nontrivial solutions and (23) represents the general solution for an arbitrary choice of x_1, x_2. In general, we observe that premature stoppage of the inversion process is equivalent to the existence of nontrivial solutions of the homogeneous system of equations. Or, in other words, *if the homogeneous system of equations possesses only the trivial solution the inversion process can be carried out to the end.*

Not all inhomogeneous systems of linear equations are solvable. For example, consider

$$2x_1 - x_2 + 3x_3 = 1$$
$$x_1 + 2x_2 - x_3 = 2$$
$$-3x_1 - x_2 - 2x_3 = 0$$

The addition of these three equations leads to the senseless statement $0 = 3$. Hence, there are no values of the unknowns which satisfy all 3 equations. This results in the desirability of a criterion for the solvability of linear equations. A sufficient criterion is given in the

Fundamental Theorem of Linear Algebra. *Consider a system of n equations with n unknowns. This system will be solvable and its solution uniquely determined if the corresponding homogeneous system possesses only the trivial solution.*

Let us first prove the existence of the solution. As we just noted, the hypothesis assures that for the matrix of the system of equations the inversion process can be carried out to the end. Hence, the inverse linear forms can be determined and a solution of the given system found by substituting the values of the right-hand sides.

To prove the uniqueness of the solution, let the given system have the form

$$\sum_{k=1}^{n} a_{ik}x_k = c_i, \qquad i = 1, 2, \ldots, n.$$

Assuming that there are two solutions $x_k^{(1)}$ and $x_k^{(2)}$, i.e., that

$$\sum_{k=1}^{n} a_{ik}x_k^{(1)} = c_i \qquad \text{and} \qquad \sum_{k=1}^{n} a_{ik}x_k^{(2)} = c_i,$$

we obtain by subtraction

$$\sum_{k=1}^{n} a_{ik}(x_k^{(1)} - x_k^{(2)}) = 0, \qquad i = 1, 2, \ldots, n.$$

Hence, the differences satisfy the homogeneous system. Since by hypothesis the latter possesses only the trivial solution, we find that

$$x_k^{(1)} - x_k^{(2)} = 0, \qquad \text{i.e.,} \qquad x_k^{(1)} = x_k^{(2)}, \qquad k = 1, 2, \ldots, n.$$

The two solutions are therefore identical. This proves the theorem.

The fundamental theorem can be generalized considerably. In fact, it holds for most linear problems in mathematics (boundary-value problems for linear differential equations, linear integral equations, etc.). In each case the existence of the solution of the given problem is assured if the corresponding homogeneous problem possesses only the zero-solution.

1.23 Considerations for the Practical Computation

While performing the exchange processes, it is forbidden to select a pivot which equals zero, and it is therefore dangerous for the accuracy of the numerical calculations to choose a pivot which is very small. This remark is simply an application of the following general principle:

When an algorithm fails for certain special exceptional values of the variables involved, then it will become inaccurate if the values of these variables are close to the exceptional values.

This principle should be observed in every numerical calculation.

In order to safeguard against inaccuracies during EX-calculations, one should always select the largest one in absolute value from among the candidates for the pivot. This rule was followed in Example 1, Appendix I.

If, in the course of an inversion calculation, a matrix is obtained which has the property that all candidates for the next pivot are small in comparison with the other elements, then we are close to an exceptional case of the type (22). In that case, the inversion is not completely stopped, but it will no longer be accurate in its further course. When we are concerned with the solution of linear equations, such a near-singular case indicates that the unknowns are only poorly determined by the given equations. One must then start looking for better equations.

1.3 Practical Solution of Linear Equations

1.31 Gauss elimination

It is also possible to solve the equations (19) by using the *inhomogeneous* linear functions

$$y_1 = 3x_1 + 5x_2 + x_3 + 4$$
$$y_2 = 2x_1 + 4x_2 + 5x_3 - 9 \tag{24}$$
$$y_3 = x_1 + 2x_2 + 2x_3 - 3$$

which we shall write in the table-form:

	x_1	x_2	x_3	1
$y_1 =$	3	5	1	4
$y_2 =$	2	4	5	-9
$y_3 =$	1	2	2	-3

$$\tag{25}$$

The linear equations require finding those values of x for which the y-variables become zero. It will not be necessary to derive the rules of an EX-step for inhomogeneous linear forms all over again; the column of the constants simply has to be transformed according to Rules 3, 4 (or 4') in 1.1. (In order to see this, we might imagine that this column belongs to a variable x_4, which always has the constant value 1.) We again try to bring the x to the upper border and the y to the left one, thereby selecting the same pivots as in (15) and (16). The first step is

	x_1	x_2	x_3	1
$y_1 =$	3	5	1	4
$y_2 =$	2	4	5	-9
$y_3 =$	1	2	2	-3
	-0.5		-1	1.5

	x_1	y_3	x_3	1
$y_1 =$	0.5	2.5	-4	11.5
$y_2 =$	0	2	1	-3
$x_2 =$	-0.5	0.5	-1	1.5

However, it is not necessary to carry along the column y_3 in the result on the right, since y_3 is to become zero. Hence, our first EX-step will look as follows:

	x_1	$\underline{x_2}$	x_3	1			x_1	x_3	1
$y_1 =$	3	5	1	4		$y_1 =$	0.5	-4	11.5
$y_2 =$	2	4	5	-9		$y_2 =$	0	1	-3
$y_3 =$	$\underline{1}$	$\underline{\underline{2}}$	$\underline{2}$	$\underline{-3}$		$x_2 =$	-0.5	-1	1.5

$$x_2 = -0.5 \qquad -1 \qquad 1.5$$

For future purposes, the basement row on the left is marked with x_2, since it corresponds exactly to the x_2-row in the right-hand scheme. If one were to continue the algorithm in this form, thereby moving another x to the left, etc., one would obtain a single column which would contain the values of the unknowns. We do not want to describe this *Jordan* elimination method in any more detail here. Instead we shall now leave off the x_2-row on the right as well. This in turn leads to the Gauss algorithm which is therefore characterized by the fact that the variables involved in the exchange are no longer carried in the new scheme. With this, the first EX-step assumes the definitive form:

	x_1	$\underline{x_2}$	x_3	1		x_1	x_3	1
	3	5	1	4		0.5	-4	11.5
	2	4	5	-9		0	1	-3
	$\underline{1}$	$\underline{\underline{2}}$	$\underline{2}$	$\underline{-3}$				

$$x_2 = \qquad -0.5 \qquad -1 \qquad 1.5$$

It might be well to go over the rules for this calculation again.

1. The basement row is obtained by dividing the pivot row by the pivot element and reversing the signs.

2. Any element of the given scheme (outside of the pivot row or the pivot column) is transformed into the corresponding element of the new scheme by adding to it the product formed with the element beneath it in the basement row as first factor, and that element of the pivot column as second factor which appears in the same row as the original element.

This EX-step can be given a different meaning and can thereby be derived in a new way by proceeding as follows:

The original equations have the form

$$3x_1 + 5x_2 + x_3 + 4 = 0$$
$$2x_1 + 4x_2 + 5x_3 - 9 = 0$$
$$x_1 + 2x_2 + 2x_2 - 3 = 0$$

By calculating x_2 from the third equation (pivot equation), we obtain the basement row

$$x_2 = -0.5x_1 - x_3 + 1.5$$

Substituting this in the first equation, we find that

$$[3 + 5(-0.5)]x_1 + [1 + 5(-1)]x_3 + [4 + 5 \cdot 1.5] = 0 \tag{26}$$

or

$$0.5x_1 - 4x_3 + 11.5 = 0$$

This is the first equation of the new scheme and (26) illustrates our second rule.

The next step in the Gauss algorithm is:

$\underline{x_1}$	x_3	1
0.5	-4	11.5
0	1	-3

x_3	1
1	-3

$$x_1 = \qquad 8 \qquad -23$$

The scheme on the right states that $x_3 - 3 = 0$, i.e., that $x_3 = 3$. The remaining unknowns are obtained by substitution in the basement rows; namely,

$$\underline{x_1} = 8x_3 - 23 = \underline{1}$$
$$\underline{x_2} = -0.5x_1 - x_3 + 1.5 = \underline{-2}$$

The following point should be noted concerning the calculation of the number of operations: While executing the algorithm for n equations, there appear, on the left of the vertical line, square matrices which successively have the following number of rows:

$$(n - 1), (n - 2), \dots, 2, 1.$$

The calculation of their elements requires

$$(n - 1)^2 + (n - 2)^2 + \dots + 2^2 + 1^2 = \frac{n^3}{3} - \frac{n^2}{2} + \frac{n}{6}$$

essential operations, and for larger n this can be approximated by $n^3/3$. The remaining calculations (basement rows, constant columns, back-substitution at the end) only produce contributions which are proportional to n^2 or n and which can, for that reason, be disregarded. Hence, the Gauss algorithm solves the equations with a computational effort about three times smaller than that of the inversion method which, as we remember, requires n^3 operations.

From the practical standpoint therefore, the *inversion method* for solving linear equations must be rejected, even in the case when the same system with several right-hand sides has to be solved. This method can be recommended only if the inverse matrix is needed for some other reasons.

For the Gauss elimination process the *sum controls* described in 1.11 have to be modified. Here the coefficient of the control variable σ is to be selected in such a way that the row sums become zero. This means that each variable x is given the value 1, and each variable y the value zero. This assures, in turn, that the process of cancelling the y-columns does not falsify the sum controls (see Appendix I, Example 2).

1.32   Subsequent Calculation of the Inverse Matrix

Consider 3 linear forms and their inverses:

$$
\left. \begin{aligned}
y_1 &= a_{11}x_1 + a_{12}x_2 + a_{13}x_3 \\
y_2 &= a_{21}x_1 + a_{22}x_2 + a_{23}x_3 \\
y_3 &= a_{31}x_1 + a_{32}x_2 + a_{33}x_3
\end{aligned} \right|
\left. \begin{aligned}
x_1 &= q_{11}y_1 + q_{12}y_2 + q_{13}y_3 \\
x_2 &= q_{21}y_1 + q_{22}y_2 + q_{23}y_3 \\
x_3 &= q_{31}y_1 + q_{32}y_2 + q_{33}y_3
\end{aligned} \right|
\tag{27}
$$

Substitution of $y_1 = 1$, $y_2 = 0$, $y_3 = 0$ in the relations on the right yields

$$
x_1 = q_{11}, \qquad x_2 = q_{21}, \qquad x_3 = q_{31}.
$$

The equations on the left then assume the form

$$
\left. \begin{aligned}
a_{11}q_{11} + a_{12}q_{21} + a_{13}q_{31} &= 1 \\
a_{21}q_{11} + a_{22}q_{21} + a_{23}q_{31} &= 0 \\
a_{31}q_{11} + a_{32}q_{21} + a_{33}q_{31} &= 0
\end{aligned} \right|
\tag{28}
$$

Hence, the first column q_{11}, q_{21}, q_{31} of the inverse matrix can be found by solving a linear system of equations. Its matrix is the matrix of the given forms and the right-hand sides are 1, 0, 0, respectively. In the same way, the second and third columns of the inverse matrix are obtained if 0, 1, 0 and 0, 0, 1, respectively, are taken as the right-hand sides. To execute the Gauss algorithm, we begin with the following scheme:

$$
\begin{array}{ccc}
x_1 & x_2 & x_3 \\
\end{array}
$$

a_{11}	a_{12}	a_{13}	-1	0	0
a_{21}	a_{22}	a_{23}	0	-1	0
a_{31}	a_{32}	a_{33}	0	0	-1

$$(29)$$

This method of inversion is advantageous whenever a system of linear equations has been solved, and when, afterwards, the matrix of the system is still to be inverted. In fact, as far as the matrix on the left in (29) is concerned, the Gauss algorithm is available from the very beginning of the inversion.

1.33   Round-off Errors

Every numerical calculation involving multiplications or divisions with a limited number of places is subject to round-off errors. For this reason, after solving a system of linear equations

$$\sum_{k=1}^{n} a_{ik}x_k + c_i = 0, \qquad i = 1, 2, \ldots, n, \tag{30}$$

by the Gauss algorithm, the values $x_1^0, x_2^0, \ldots, x_n^0$ of the unknowns, should, for control purposes, be substituted back into the equations. The right-hand sides in (30) will not turn out to be exactly zero, but small values r_i, called residuals, will appear:

$$\sum_{k=1}^{n} a_{ik}x_k^0 + c_i = r_i. \tag{31}$$

To improve the solutions, the corrections

$$\Delta x_k = x_k - x_k^0$$

are introduced. Subtraction of (31) from (30) results in the following system of linear equations for the corrections:

$$\sum_{k=1}^{n} a_{ik}\Delta x_k + r_i = 0. \tag{32}$$

This system has the same matrix as the original one (30), so that the Gauss algorithm will have to be repeated only in the column of the constants.

Moreover, calculation with a smaller number of digits suffices, since both the residuals and the corrections are small quantities. This procedure illustrates the general *principle of correction* in numerical analysis. In order to calculate the desired quantities, it is often better not to look for an algorithm which gives these quantities directly. Rather, it is frequently more advantageous to proceed from some approximate values and to develop only methods for the calculation of the corrections. Such correction algorithms are often simpler than direct algorithms and, for the most part, can be executed with a smaller number of digits.

1.34 Dominant Diagonal, Compact Algorithm

The principle of selecting the largest in absolute value from among all qualifying pivots should be observed with the Gauss algorithm as well. However, one would occasionally like to prescribe the pivot selection from the very beginning for the purpose of further schematizing the computing process.

This is permissible only when—for some reason—it is known that no small, or even vanishing, pivots will occur. We shall now discuss a case of practical importance for which fixed-sequence pivot selection is permitted and, in fact, recommended. Let us assume that in the given matrix

$$
\begin{array}{ccc}
\underline{a_{11}} & a_{12} & a_{13} \\
a_{21} & \underline{a_{22}} & a_{23} \\
a_{31} & a_{32} & \underline{a_{33}}
\end{array}
\tag{33}
$$

the elements on the diagonal (from the upper left to the lower right) heavily predominate over the other matrix elements, i.e., that their absolute values are substantially greater than those of the other, off-diagonal elements. This assumption permits selection of the element in the upper left-hand corner as pivot. According to the rectangle rule in 1.1, the first step of the Gauss elimination process then produces the 2×2 matrix

$$
\begin{array}{cc}
a_{22} - \dfrac{a_{12}a_{21}}{a_{11}} & a_{23} - \dfrac{a_{13}a_{21}}{a_{11}} \\[3mm]
a_{32} - \dfrac{a_{12}a_{31}}{a_{11}} & a_{33} - \dfrac{a_{13}a_{31}}{a_{11}}
\end{array}
\tag{34}
$$

Hence, during the transition from (33) to (34), the matrix elements undergo corrections which have been formed with the off-diagonal elements and are therefore small. Consequently, the diagonal in (34) is again dominant and the upper left-hand element can again be selected as pivot. The same pivot selection applies to the further elimination steps.

The assumption we made here and the conclusions drawn from it are at most plausible. Furthermore, the dominance of the diagonal elements can be sharply reduced during the Gauss elimination process. Nevertheless, in Chapter 3 (on least-squares) we shall give in a special case a wholly precise presentation of these observations.

As an example, let us solve the following system of 4 equations:

	x_1	x_2	x_3	x_4	1	
	a_{11}	a_{12}	a_{13}	a_{14}	c_1	
	a_{21}	a_{22}	a_{23}	a_{24}	c_2	(35)
	a_{31}	a_{32}	a_{33}	a_{34}	c_3	
	a_{41}	a_{42}	a_{43}	a_{44}	c_4	
$x_2 =$		α_{12}	α_{13}	α_{14}	γ_1	

i.e., the first equation reads

$$a_{11}x_1 + a_{12}x_2 + a_{13}x_3 + a_{14}x_4 + c_1 = 0.$$

We also added the basement row. Assume that the Gauss elimination leads successively to the following tables:

(36)

x_2	x_3	x_4	1
a'_{22}	a'_{33}	a'_{24}	c'_2
a'_{32}	a'_{33}	a'_{34}	c'_3
a'_{42}	a'_{43}	a'_{44}	c'_4
$x_2 =$ α'_{23} α'_{24} γ'_2

x_3	x_4	1
a''_{33}	a''_{34}	c''_3
a''_{43}	a''_{44}	c''_4
$x_3 =$ α''_{34} γ''_3

x_4	1
a'''_{44}	c'''_4
$x_4 =$ γ'''_4

In order to show that considerable writing effort can be saved, we place the pivot columns (first columns) and the basement rows together in a new table and use this table to develop a modification of the algorithm:

a_{11}	α_{12}	α_{13}	α_{14}	γ_1	$= x_1$
a_{21}	a'_{22}	α'_{23}	α'_{24}	γ'_2	$= x_2$
a_{31}	a'_{32}	a''_{33}	α''_{34}	γ''_3	$= x_3$
a_{41}	a'_{42}	a''_{43}	a'''_{44}	γ'''_4	$= x_4$

$$(37)$$

We assert that this table can be calculated directly from the given table (35) without execution of the intermediate steps (36). We begin with an element below the staircase lines. Application of the Gauss elimination rule then yields, for example,

$$a'_{32} = a_{32} + a_{31}\alpha_{12}$$

or

$$a''_{43} = a'_{43} + a'_{42}\alpha'_{23} = (a_{43} + a_{41}\alpha_{13}) + a'_{42}\alpha'_{23}$$

Similarly, for elements enclosed within the staircase we obtain, for example,

$$a'''_{44} = a''_{44} + a''_{43}\alpha''_{34} = (a'_{44} + a'_{42}\alpha'_{24}) + a''_{43}\alpha''_{34}$$
$$= a_{44} + a_{41}\alpha_{14} + a'_{42}\alpha'_{24} + a''_{43}\alpha''_{34}$$

In order to express these results verbally, we shall use the following mathematical term: By the *scalar product* of two given rows of numbers.

$$p_1, \ p_2, \ p_3, \ \cdots$$
$$q_1, \ q_2, \ q_3, \ \cdots$$

we mean the construction

$$p_1q_1 + p_2q_2 + p_3q_3 + \cdots$$

Our calculations can then obviously be crystallized in the rule:

1.   An element below or within the staircase of the new table (37) is obtained as follows: Form the scalar product of the row of numbers to the left of it with the column of numbers above it in this table and then add to the result the corresponding element in the original table (35). (All rows and columns of numbers mentioned here begin at the left or upper border, respectively.)

As an example for elements above the staircase, we use

$$\alpha''_{34} = -\frac{a''_{34}}{a''_{33}} = -\frac{1}{a''_{33}}(a'_{34} + a'_{32}\alpha'_{24}) = -\frac{1}{a''_{33}}(a_{34} + a_{31}\alpha_{14} + a'_{32}\alpha'_{24})$$

$$\gamma''_3 = -\frac{c''_3}{a''_{33}} = -\frac{1}{a''_{33}}(c'_3 + a'_{32}\gamma'_2) = -\frac{1}{a''_{33}}(c_3 + a_{31}\gamma_1 + a'_{32}\gamma'_2).$$

This brings us to the following rule:

2. An element above the staircase of the new table (37) is obtained by applying, first of all, the above Rule 1. The result is then to be divided by the diagonal element to the left of it and the sign reversed.

To calculate the new table (37), we start with the first column [which is identical with the first column of (35)] and with the first row. The latter is produced by dividing elements of the first row of (35) by $(-a_{11})$. Then the calculations are continued by using Rules 1 and 2 and by proceeding in the same sequence as that in which the lines of a book are read. As before, the unknowns are found by back-substitution in the basement rows, which, of course, are available in (37).

For this compact algorithm, the number of essential operations is the same as for the normal Gauss algorithm. When the computations are performed on a desk-calculator, every element of the new table (37) can be written down directly without need of noting down intermediate results. (Appendix I, Example 3.)

In the case of a dominant diagonal an approximate solution of the correction equations (32)

$$\sum_{k=1}^{n} a_{ik}\Delta x_k + r_i = 0$$

for the improvement of any approximation values x_k^0 can be obtained by neglecting completely the off-diagonal elements of the matrix. In this way, we find that

$$\Delta x_i = -\frac{r_i}{a_{ii}}. \tag{38}$$

If the improved solutions $(x_i^0 + \Delta x_i)$ are again substituted in the given equations, new residuals result which once more can be corrected by means of (38). In this way, an iterative procedure for solving the system of equations has been started which gives increasingly better approximations of the solutions. However, we must not be tempted to use such methods of successive approximations in cases where the diagonal is only weakly or not at all dominant. A thorough treatment of iteration methods can be found in [1], [2], [3] (see the list of references).

2 Linear Programming

This branch of linear algebra is concerned with extremum problems. A linear function of several variables is to be maximized (or minimized) under constraints which also have a linear character. Linear programming is the most important mathematical basis of "operations research." This science concerns itself, for example, with the problem of designing the organization of a manufacturing process optimally in such a way that the cost of the product, for instance, is minimal. We will have to limit ourselves here to a presentation of the mathematical principles; for their extensive fields of application, we refer to the special literature [4], [5], [34][1]. In the following, we assume a knowledge of the material in 1.1.

2.1 An Introductory Example

A farmer owns 100 acres of land. He wants to plant potatoes in one part, corn in another and perhaps leave the third section fallow. In addition to this, we have the following information:

	Potatoes	Corn	Total available
Cultivation costs in dollars per acre	10	20	1100
Workdays per acre	1	4	160
Net profit in dollars per acre	40	120	

The figures in the last column indicate that the farmer has a capital of $1100 and 160 workdays to spend. The question now is: How is he to organize his planting so as to realize a maximum net profit?

If x_1 acres are planted with potatoes and x_2 acres with corn, the first two rows of the table state that

$$10x_1 + 20x_2 \leq 1100, \qquad x_1 + 4x_2 \leq 160. \tag{39}$$

[1] The formal algebraic relations between the exchange method and the computational method of linear programming have been pointed out by A. W. Tucker, who has also developed a general theory of the subject. (Compare A. W. Tucker in *Proceedings of Symposia in Appl. Math.*, **X**, 129–140, 1960.)

Additionally we have

$$x_1 + x_2 \leq 100 \qquad (40)$$

and

$$x_1 \geq 0, \qquad x_2 \geq 0. \qquad (41)$$

The unknowns x_1 and x_2 are thus bound by five *linear inequalities*. The problem of maximizing the net profit can be formulated as follows:

$$40x_1 + 120x_2 = \text{max.} \qquad (42)$$

In order to write the inequalities uniformly, it is best to introduce the linear functions:

$$y_1 = -10x_1 - 20x_2 + 1100, \qquad y_2 = -x_1 - 4x_2 + 160,$$
$$y_3 = -x_1 - x_2 + 100. \qquad (43)$$

The five inequalities (39) through (41) then simply state that both the independent variables x_1 and x_2 and the dependent variables y_1, y_2, y_3 must have positive values. In all our considerations of linear programming, a linear constraint will always be formulated so as to require a linear function to be positive. Furthermore, we introduce the *objective function*

$$z = 40x_1 + 120x_2$$

which is to be maximized. Following the methods of 1.1., we tabulate all functions, and we shall choose here the *transposed notation* as described in 1.12:

	$y_1 =$	$y_2 =$	$y_3 =$	$z =$
$-x_1$	10	1	1	-40
$-x_2$	20	4	1	-120
1	1100	160	100	0

(44)

(Table of the linear program)

2.11 Graphical Solution

A linear program with 2 unknowns x_1, x_2 can be easily solved by graphical means (Fig. 1). In Cartesian coordinates x_1, x_2 the equation $y_1 = 0$, i.e., $-10x_1 - 20x_2 + 1100 = 0$, represents a straight line. This line has been drawn into the figure and denoted by y_1. Now, if a point (x_1, x_2) is to satisfy our condition $y_1 \geq 0$, it has to be in that half plane which is bounded by this straight line and contains the origin (since the origin satisfies the inequality). Similarly, the two conditions $y_2 \geq 0$ and $y_3 \geq 0$ yield two further half planes; and finally, $x_1 \geq 0$ and $x_2 \geq 0$ define the half planes to the right of the x_2-axis and above the x_1-axis, respectively. A point (x_1, x_2) which satisfies all five conditions must lie in that area which these five half planes have in common. This is the *convex polygon* shaded in the figure. In our problem, only the points within and on the border of this polygon are *feasible*.

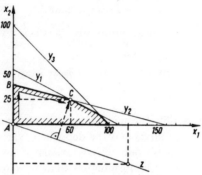

FIG. 1.   Linear program in two variables.

The function z is proportional to the distance between the point (x_1, x_2) and the straight line $40x_1 + 120x_2 = 0$ (in Fig. 1, also denoted by z). This fact is easily ascertained by drawing some contour lines $z = $ const of this function; they are straight lines parallel to z. Hence, our extremum problem can be interpreted geometrically in the following way: *among all points inside or on the boundary of the convex polygon that point is to be found which has maximal distance from the straight line z.* This geometric description yields the corner C as the solution. Its coordinates

$$x_1 = 60, \qquad x_2 = 25 \tag{45}$$

are the solutions of the linear program. Hence, the farmer should plant 60 acres of potatoes and 25 acres of corn while 15 acres of his property remain unused.

2.2 Solution of a Program with the Exchange Method

A general linear program is determined by the m linear functions of n variables $x_1, x_2, \ldots, x_n$

$$y_i = -\sum_{k=1}^{n} a_{ik}x_k + c_i, \qquad i = 1, 2, \ldots, m \tag{46}$$

and by the objective function

$$z = -\sum_{k=1}^{n} a_k x_k + c. \tag{47}$$

The problem is to maximize z by appropriate choice of x_k subject to the condition that all variables y_i and x_k have to be positive:

$$y_i \geq 0, \qquad x_k \geq 0.$$

Frequently it will also be required to minimize the objective function. This is nothing new because the problem $z = \min$ is equivalent to $(-z) = \max$.

The use of equations[1] as constraints can also be reduced to the above case. In fact, an equation $y_i = 0$ can be replaced by the two inequalities $y_i \geq 0$ and $(-y_i) \geq 0$.

The formal scheme of this linear program has the form

$$
\begin{array}{r|ccc|c}
 & \cdots\, \underline{y_i =} & \cdots\, \underline{y_j =} & & \cdots\, \underline{z =} \\
\hline
 & \vdots & \vdots & & \vdots \\
\underline{-x_k} & \cdots\, \underline{\underline{a_{ik}}} & \cdots\, \underline{a_{jk}} & \cdots & \underline{a_k} \\
 & \vdots & \vdots & & \vdots \\
\hline
1 & \cdots\, \underline{c_i} & \cdots\, \underline{c_j} & \cdots & c
\end{array}
\tag{48}
$$

For the following method of solution, we must assume that all elements in the last row, except for the element in the lower right corner, are positive:

$$c_i > 0, \qquad i = 1, 2, \ldots, m.$$

We shall then proceed with an exchange-step as defined in 1.1 which satisfies the following two requirements:

1st. *Requirement.* In the new scheme, the elements in the last row are again to be > 0.

2nd. *Requirement.* The element in the lower right corner is to increase (this is meant in the "strict" sense, i.e., the element is not to remain constant!).

We are not in a position to justify these requirements here, since they are founded on a geometric interpretation of the method to be described next,

---

[1] However, usually one deals with such constraints by solving them for the unknowns. For example, consider the conditions

$$3x_1 + 2x_2 + x_3 - 4x_4 = 1$$
$$x_1 + x_2 - 2x_3 + 3x_4 = 1, \qquad x_i \geq 0.$$

By solving them with respect to x_1 and x_2, we obtain

$$x_1 = -5x_2 + 10x_4 - 1 \geq 0$$
$$x_2 = 7x_3 - 13x_4 + 2 \geq 0.$$

Here x_3 and x_4 are now the independent variables, while x_1 and x_2 are linear functions which can be treated in the same way as (46). In more complicated cases, such solution processes can be carried out with the help of the exchange method.

and this interpretation assumes a knowledge of n-dimensional geometry. However, we shall give this geometric illustration for simpler cases later.

According to 1.12, we have to execute an EX-step in exact accordance with Rules 1 through 4 of 1.1, even if the scheme is given in the transposed notation. Hence, if in table (48) the variables x_k and y_i are to be exchanged (pivot a_{ik}), the elements of the scheme will be transformed as follows:

$$c_i \to \frac{c_i}{a_{ik}}, \qquad c_j \to c_j - \frac{a_{jk}}{a_{ik}} c_i, \qquad c \to c - \frac{c_i a_k}{a_{ik}}. \tag{49}$$

From the first relation and the first requirement mentioned above, we immediately conclude that $a_{ik} > 0$; hence, the pivot element has to be strictly positive. But the third relation and the second requirement then imply that $a_k < 0$. In other words, the pivot row must be chosen in such a way that its last element is negative. Finally, we obtain from the first requirement and the second relation (49) that

$$c_j - \frac{a_{jk}}{a_{ik}} c_i > 0$$

for fixed i and all $j \neq i$. This condition connects the elements of the pivot row and those of the last row of (48). It is automatically satisfied if $a_{jk} \leq 0$. Otherwise, we can divide by a_{jk} and obtain

$$\frac{c_j}{a_{jk}} > \frac{c_i}{a_{ik}}, \qquad j = 1, 2, \dots, m, \qquad \text{but} \qquad j \neq i. \tag{50}$$

Evidently the quotients obtained by dividing every element in the last row by the element above it in the pivot row play an important role. We call these the *characteristic quotients*. The inequalities (50) then state: The quotient belonging to the pivot element has to be minimal among all those characteristic quotients formed with positive elements of the pivot row as denominators. In conclusion, we obtain the following rules:

1. *Rule for the choice of the pivot row.* The element in the last column of the pivot row has to be < 0.

2. *Rule for the choice of the pivot in the pivot row.* Select all elements of the pivot row which are > 0 and form the corresponding characteristic quotients. The smallest one among these determines the position of the pivot.[1]

---

[1] Here we must make a note of the following exceptional case. It may happen that two (or more) elements of the last row yield the minimal value of the characteristic quotients and hence that several elements of the pivot row qualify as pivot. Choosing one of these and executing the EX-step, the elements in the last row of the new table will still be ≥ 0, but some of them may be $= 0$. Hence, the first requirement is not strictly satisfied. In general, we shall call a table degenerate if there are any zeros in the last row. This phenomenon of degeneration will be discussed in 2.3. During a

Let us now treat the problem of our farmer in this new way:

	$y_1 =$	$y_2 =$	$y_3 =$	$z =$		$y_1 =$	$x_2 =$	$y_3 =$	$z =$
$-x_1$	10	1	1	-40	$-x_1$	5	0.25	0.75	-10
$-x_2$	20	4	1	-120	$-y_2$	-5	0.25	-0.25	30
1	1100	160	100	0	1	300	40	60	4800
	55	40	100			60	160	80	

$$(51)$$

On the left-hand side, we have again provided the given program. According to Rule 1, both rows above the dividing line qualify as pivot rows. We have selected the lower one and noted down the corresponding characteristic quotients, namely, 55, 40 and 100. The smallest among these is $= 40$ and hence, according to Rule 2, it identifies the second column as the pivot column. The exchange was calculated according to Rules 1 through 4 of 1.1 and the results are given in the table on the right-hand side. In it, all our requirements are satisfied; the elements in the last row (on the left of the dividing line) are positive and the element in the lower right-hand corner increased from zero to 4800.

For the next exchange, only the first row qualifies as pivot row. Its first element proves to have the smallest characteristic quotient and is therefore the pivot. The execution of the EX-step yields

	$x_1 =$	$x_2 =$	$y_3 =$	$z =$
$-y_1$	*	*	*	2
$-y_2$	*	*	*	20
1	60	25	15	5400

$$(52)$$

We calculated only the elements on the border of the table inasmuch as the algorithm now stops, since all elements of its last column have become positive (Rule 1). In particular, it follows from this last scheme that

$$z = -2y_1 - 20y_2 + 5400.$$

practical calculation, one need not concern oneself with this but can proceed with the calculation according to the above rules—also when the table is degenerate. In this case, however, it can happen that the element in the lower right-hand corner does not increase but remains constant; i.e., that such an EX-step does not strictly satisfy the second requirement

The condition $y_1 \geq 0$, $y_2 \geq 0$ therefore implies that the objective function z is always ≤ 5400. This upper bound is attained for the choice of $y_1 = 0$, $y_2 = 0$. This choice is permissible since (52) then yields

$$\underline{\underline{x_1 = 60}}, \qquad \underline{\underline{x_2 = 25}}, \qquad y_3 = 15. \tag{53}$$

These values turn out to be positive due to the fact that our first requirement was satisfied in scheme (52) as well. Therefore, all five constraints

$$x_1 \geq 0, \qquad x_2 \geq 0, \qquad y_1 \geq 0, \qquad y_2 \geq 0, \qquad y_3 \geq 0$$

are satisfied. Hence, the maximum of z is $z = 5400$ (the farmer's net profit); according to (53), it is attained for $x_1 = 60$ and $x_2 = 25$, which bears out the graphical solution in 2.11.

This discussion of table (52) provides a fairly general understanding of the two requirements formulated above.

Our algorithm therefore consists of an (*a priori* undetermined) number of EX-steps to be executed in accordance with the above two rules for the choice of the pivot. It is also called the *simplex algorithm*. Even in the case of a general program (46), this algorithm leads to the desired solution in a finite number of steps, provided no degenerate table can appear. In fact, the $(n + m)$ variables x_k, y_i can be written only in a finite number of ways along the border of a rectangle of the form (48). Hence, *a priori*, there are only finitely many tables which can occur at all. Moreover, during the simplex algorithm, the same table can never appear twice because, with each step, the element in the lower right-hand corner must increase. The algorithm therefore has to stop after finitely many steps. This can happen in two ways:

1) All elements of the z column are ≥ 0 (Rule 1 fails). A consideration analogous to the one in the example shows that the desired maximum has been obtained.

2) Although there are still negative elements in the z-column, the elements in the corresponding rows are without exception ≤ 0 (Rule 2 fails). Setting those variables on the left-hand border equal to zero which correspond to the remaining rows, one observes that in this case z can be arbitrarily large, while all constraints are satisfied. This means that the given problem has no solution at all.

2.3 Geometric Interpretation of the Simplex Algorithm

As a result of our discussion of the farm problem we had to set the variables y_1, y_2, i.e., the variables on the left-hand border of table (52), equal to zero. Let us also set the left-hand variables equal to zero in the previous tables (51). As far as the table on the left-hand side of (51) is concerned, setting $x_1 = 0$

and $x_2 = 0$ causes the other variables to have the values

$$y_1 = 1100, \qquad y_2 = 160, \qquad y_3 = 100,$$

and, by assumption, these are positive. All five constraints are again satisfied. In Fig. 1, this means that the point $x_1 = 0$, $x_2 = 0$ (i.e., the origin of the coordinate system) is a feasible point. In other words, it belongs to the convex polygon. Moreover, this point is a corner A of the polygon since it is the intersection of the two sides $x_1 = 0$ and $x_2 = 0$. Hence, the scheme on the left-hand side of (51) corresponds to this corner.

A similar consideration for the table on the right-hand side yields the point

$$x_1 = 0, \qquad y_2 = 0,$$

i.e., the intersection of the sides $x_1 = 0$ and $y_2 = 0$ of the polygon, or the corner B; there $z = 4800$.

The variable x_1 occurs on the left-hand border of both tables (51), and disappears in A as well as in B. Hence we can visualize the first EX-step to mean that we move along the side $x_1 = 0$ from corner A to corner B.

During the second EX-step leading from (51) to (52), we have the joint side $y_2 = 0$ along which we move to the final corner C. In C we then also have $y_1 = 0$.

In conclusion, we can say that the simplex method consists in moving through a continuous *chain of sides* of the convex polygon. In Fig. 1, this has been indicated by arrows.

For each corner of the chain of sides, i.e., for the corners A, B, C, we have a corresponding scheme and in it the corresponding value of the objective function always appears in the lower right-hand corner. This value increases with each EX-step and hence *the objective function also increases while we move through the chain of sides* until it attains its maximal value in C.

Further information is obtained when a linear program (46) with three independent variables x_1, x_2, x_3 ($n = 3$, m arbitrary) is represented geometrically in a three-dimensional coordinate system. The constraints $y_i \geq 0$, $x_k \geq 0$ then determine a *convex polyhedron*, and the assumption $c_i > 0$ again states that the origin is a corner A of the polyhedron. The linear program requires locating that point of the polyhedron for which the distance from the plane

$$z = -\sum_{k=1}^{3} a_k x_k + c = 0 \tag{54}$$

is maximal. In the following discussion, it will be useful to rotate the entire configuration in the space in such a way that the plane (54) will be horizontal. Our problem then can be put simply: *Given a convex polyhedron in three-dimensional space and one of its corners A, find its highest point.* Figure 2 attempts to show such a situation in perspective. Again the simplex algorithm solves this problem by constructing a continuous train of edges $A\ B\ C\ D\ E$ which leads constantly upward until it attains the highest point.

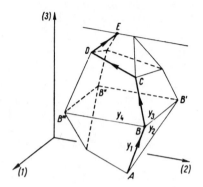

FIG. 2.   Linear program in three variables.

Figure 2 furthermore gives us a chance to discuss some special occurrences. In general, a corner of the polyhedron will constitute the intersection of three of its side planes.   If (as, for example, in point B) four planes pass through the same corner, this may be considered a coincidence and we shall call such a corner degenerate.   We will now determine what its corresponding table looks like.   According to the notation in Fig. 2, the first edge $A\ B$ of the chain of edges is characterized by $y_1 = 0$, $y_2 = 0$.   In point B, we additionally have the equation $y_3 = 0$, so that in the table corresponding to B, three variables occupy the left-hand border:

	$\dots y_4 = \dots$	$z =$
$-y_1$	*	0
$-y_2$	$+$	$-$
$-y_3$	*	$+$
1	0	

We now have $y_4 = 0$ in corner B as well.   The lowest element in the y_4-column will therefore have to be zero and so we have a degenerate table (cf. footnote in 2.2).   In other words, a degenerate corner corresponds to a degenerate table and vice versa.   In Fig. 2, we assumed that the edge $B\ B'$ (characterized by $y_2 = 0$, $y_3 = 0$) was horizontal and hence that the z-element in the first row equalled zero.   In view of Rule 1 of 2.2, this row certainly cannot become the pivot row for the next step of the simplex algorithm. Our chain of edges can therefore not continue along $B\,B'$, and we avoid the danger of running through the cycle $B\,B'\,B''\,B'''$ of the four horizontal edges which leads back to corner B. Otherwise it would simply mean wasted effort.

Since the edge $B\,A$ leads downward, neither y_1 nor y_3 will be exchanged, and, for this reason, y_2 will have to be exchanged. In view of the fact that the remaining planes $y_1 = 0$, $y_3 = 0$ do not form an edge of the polyhedron, this will be possible only if y_2 is exchanged with y_4. In fact, following the EX-step we then obtain $y_1 = y_3 = y_4 = 0$ or, in other words, we remain in corner B ("*stationary exchange*"). We can also support this result by performing a simple check which ascertains that the signs in the above table are correct. The rules of the simplex algorithm accordingly give the unique result that y_2 and y_4 must be exchanged.

After the stationary exchange, corner B is characterized by $y_1 = y_3 = y_4 = 0$ and the edge $y_3 = y_4 = 0$ is available for continuation toward C. In the example given in Fig. 2, the degeneracy, therefore, does not disturb anything. However, in more complex cases it is possible that the simplex algorithm leaves us in one corner. This is the case if the algorithm produces a sequence of stationary EX-steps which leads back to the initial table; i.e., which forms a cycle through which we keep running. Methods have been developed which avoid this case [4], [34].

The upper edge of the polyhedron in Fig. 2 is horizontal. Each point of this edge therefore has the same value of the objective function as the end point E. Hence, each of these points is also a solution of the linear program. In other words, programs exist which have several solutions.

2.4 Generalizations

There are linear programs where it is not required that all independent variables x_k in (46) be positive. Let us discuss with the help of an example how we should proceed in such a case. As before, we will assume that the elements c_i of the last row are > 0. Consider the following program:

	$y_1 =$	$y_2 =$	$y_3 =$	$z =$	
$-x_1$	$\underline{1}$	3	0	-1	
$-x_2$	$\underline{\underline{-2}}$	$\underline{-4}$	$\underline{1}$	$\underline{1}$	(55)
1	$\underline{2}$	6	3	0	
	$\underline{-1}$	-1.5			

We assume only that $x_1 \geq 0$ and, of course, that $y_1 \geq 0$, $y_2 \geq 0$, $y_3 \geq 0$.

Since x_2 is subject to no conditions, we can simply eliminate this variable from the calculation by taking the corresponding row as pivot row. In the table obtained after this EX-step, the x_2-column will then be left out. Hence, the lowest element in this column does not have to be positive. However, the first and also the second requirement of 2.2 for the remaining elements of the new last row have to remain valid. To satisfy these requirements, we must repeat the discussion about the formulas (49) and the table (48). In our example, the last element a_k of the pivot row is positive. Hence, the second requirement, together with the third relation (40), implies that the pivot element a_{ik} must be negative. Then it follows from the first requirement and the second relation that

$$c_j - \frac{a_{jk}}{a_{ik}} c_i > 0 \qquad \text{for all} \qquad j \neq i.$$

This condition is trivially satisfied if $a_{jk} \geq 0$. Otherwise we can divide by $(-a_{jk})$ and obtain

$$\frac{c_j}{a_{jk}} < \frac{c_i}{a_{ik}}, \qquad j \neq i.$$

Both sides of this inequality are < 0. Hence, among all characteristic quotients formed with negative elements of the pivot row, we must select the greatest one (i.e., that quotient which is farthest to the right on the real axis). The first relation (49) need not be considered because it merely concerns that column which will be left out in the new scheme.

The characteristic quotients have been written down underneath the scheme in (55); the maximal one of these is (-1) and it determines that the first column has to be the pivot column. The exchange now yields

	$y_2 =$	$y_3 =$	$z =$
$-x_1$	1	0.5	-0.5
$-y_1$	-2	0.5	0.5
1	2	4	1
	2	8	

(56)

and the column which was left out had the form

$$x_2 = 0.5x_1 + 0.5y_1 - 1. \tag{57}$$

In this way, we have obtained a linear program where all the variables x_1, y_1, y_2, y_3 have to be ≥ 0. Since, by construction, the last row is positive, this program can be solved by the simplex algorithm described in 2.2.

If, in the formulation of our program (55), the constraint $x_1 \geq 0$ had not been given either, we would now have to eliminate the element x_1 as well, and the first row would become the pivot row. Since its last element is negative, Rule 2 in 2.2. applies for the choice of the pivot: The smallest characteristic quotient is $= 2$, and therefore the first column is the pivot column.

After two steps of the simplex algorithm, we obtain as solution of the program (56):

$$y_2 = 0, \qquad y_3 = 0, \qquad y_1 = 2, \qquad x_1 = 6, \qquad z = 3$$

and (57) then yields the missing "free" variable $x_2 = 3$. We shall note down the result of our operations as follows:

Rule for the elimination of a free variable. If the last element in the corresponding row of the scheme is positive (negative), choose those elements in that row which are < 0 (> 0); and form the corresponding characteristic quotients. The greatest (smallest) of these then determines the place of the pivot.[1]

2.41 The Most General Case

Up to this time, we have always assumed that in a given linear program the constants c_i have to be > 0. We shall now remove this condition. As an example, consider the program

$$x_1 + 4x_2 \geq 8, \qquad 2x_1 + 3x_2 \geq 12, \qquad 2x_1 + x_2 \geq 6$$
$$x_1 \geq 0, \qquad\qquad x_2 \geq 0, \qquad\qquad x_1 + x_2 = \min. \tag{58}$$

We introduce the functions

$$y_1 = x_1 + 4x_2 - 8, \qquad y_2 = 2x_1 + 3x_2 - 12, \qquad y_3 = 2x_1 + x_2 - 6,$$
$$z = -(x_1 + x_2) = \max. \tag{59}$$

The corresponding scheme is:

	$y_1 =$	$y_2 =$	$y_3 =$	$z =$
$-x_1$	-1	-2	-2	1
$-x_2$	-4	-3	-1	1
1	-8	-12	-6	0

Now the last row of the c_i is no longer positive. To proceed in this case, we first look for special values of the independent variables x_1, x_2 which satisfy

---

[1] If this last element is equal to zero, we may arbitrarily regard it as either positive or negative.

the five constraints in (58). Such values of x_1, x_2 shall be called a pair of "feasible values." Geometrically (Fig. 3), this means finding a feasible point or, in other words, a point which lies inside the shaded convex polygon. Immediately following this discussion, we shall describe a general method

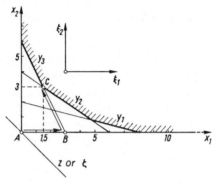

FIG. 3.   Dual solution method.

which permits finding feasible values. However, in our example (58) it suffices to make x_1, x_2 sufficiently large to satisfy the constraints. We will select $x_1 = 3$, $x_2 = 4$. The substitution

$$x_1 = \xi_1 + 3, \qquad x_2 = \xi_2 + 4$$

then introduces the chosen feasible point as the origin of the new ξ_1, ξ_2-coordinate system. After re-evaluating everything in terms of these new variables, we obtain the scheme

	$x_1 =$	$x_2 =$	$y_1 =$	$y_2 =$	$y_3 =$	$z =$
$-\xi_1$	-1	0	-1	-2	-2	1
$-\xi_2$	0	-1	-4	-3	-1	1
1	3	4	11	6	4	-7

(60)

In this program, the last row has now become positive (this is, of course, due to the fact that the values $\xi_1 = 0$, $\xi_2 = 0$, i.e., $x_1 = 3$, $x_2 = 4$, satisfy all constraints and hence make all variables x_1, x_2, y_1, y_2, y_3 positive). In the new program (60), ξ_1, ξ_2 are free variables, subject to no constraints. They therefore will be eliminated with the help of the technique developed above, after which we can apply the simplex algorithm. This yields the solution

$$x_1 = 1.5; \qquad x_2 = 3$$

which can also be obtained from Fig. 3.

Calculation of a feasible point (solution of a system of linear inequalities).
Consider a system of linear inequalities

$$\sum_{k=1}^{n} a_{ik}x_k + c_i \geq 0 \qquad (60a)$$

for the unknowns $x_1, x_2, \ldots, x_n$.

The problem is to determine whether a solution exists and, if so, how it can be calculated. (For example, such a system can consist of all the constraints of a linear program. Our problem then is to find a feasible set of values x_k which can be used as starting point for the algorithms of linear programming.) To solve this problem, we introduce the extended system

$$\sum_{k=1}^{n} a_{ik}x_k + x_{n+1} + c_i \geq 0 \qquad (60b)$$

where we have added a new unknown x_{n+1}. It is easy to find a set of values $x_1, x_2, \ldots, x_{n+1}$ which satisfy (60b). In fact, we only have to choose x_{n+1} sufficiently large. Using this set of starting values, we now solve that linear program which requires x_{n+1} to be minimal, subject to the constraints (60b). (In other words, the objective function is the last variable x_{n+1}.) Let $\xi_1, \xi_2, \ldots, \xi_{n+1}$ be a solution of the program, so that

$$\sum_{k=1}^{n} a_{ik}\xi_k + \xi_{n+1} + c_i \geq 0.$$

If $\xi_{n+1} \leq 0$, it follows that

$$\sum_{k=1}^{n} a_{ik}\xi_k + c_i \geq 0$$

and hence we have found a solution of the original system (60a).

Conversely, assume that (60a) is solvable and that $x_1^{(0)}, x_2^{(0)}, \ldots, x_n^{(0)}$ is a solution. Evidently then

$$x_1^{(0)}, \ldots, x_n^{(0)}, \qquad x_{n+1}^{(0)} = 1$$

is a solution of (60b). Therefore, in the minimal solution of our linear program ξ_{n+1} cannot be greater than zero and consequently $\xi_{n+1} \leq 0$. Or, in other words, if $\xi_{n+1} > 0$, the given system (60a) does not have a solution.

2.5 Method of Dual Solution

We have previously been using the transposed notation (1.12). Let us now develop the table of the linear program (58) in the normal notation, where we take as objective function exactly that function $\zeta = x_1 + x_2$ which has to be made minimal:

	x_1	x_2	1
$y_1 =$	1	4	-8
$y_2 =$	2	3	-12
$y_3 =$	2	1	-6
$\zeta =$	1	1	0
	0.5	1	

$$(61)$$

Since the last row is positive, the table can be reduced by the simplex algorithm. We will take a look at the results of this process with no specific aim in mind at this time, other than general interest. According to Rule 1 in 2.2, a row with a negative element in the last column must be selected as pivot row; we chose the third row. The characteristic quotients have been noted down below; according to Rule 2 in 2.2, the smallest one designates the first column as a pivot column. The first step of the simplex procedure is an exchange of the variables x_1 and y_3 and produces the following table on the left:

	y_3	x_2	1
$y_1 =$	0.5	3.5	-5
$y_2 =$	1	2	-6
$x_1 =$	0.5	-0.5	3
$\zeta =$	0.5	0.5	3
	0.5	0.25	

	y_3	y_2	1
$y_1 =$	*	*	5.5
$x_2 =$	*	*	3
$x_1 =$	*	*	1.5
$\zeta =$	0.25	0.25	4.5

$$(62)$$

The next step then produces the table on the right, which at the same time calls a halt to the entire procedure, since the last column has become positive. We will consider the last table in the same way as the problem of the farmer in 2.2. From the last row we find that

$$\zeta = 0.25y_3 + 0.25y_2 + 4.5.$$

Because $y_3 \geq 0$ and $y_2 \geq 0$, the objective function ζ is always ≥ 4.5; it assumes this lower bound for the choice $y_3 = 0$, $y_2 = 0$, and this choice is feasible since the table then yields

$$y_1 = 5.5; \qquad x_2 = 3; \qquad x_1 = 1.5.$$

These values are ≥ 0 because, after the last EX-step, the last column has become positive. Hence, all five constraints

$$x_1 \geq 0; \qquad x_2 \geq 0; \qquad y_1 \geq 0; \qquad y_2 \geq 0; \qquad y_3 \geq 0$$

are satisfied. The linear program's solution (already given in 2.41) is $x_1 = 1.5$; $x_2 = 3$, and the minimal value of the objective function is $\zeta = x_1 + x_2 = 4.5$. Since the values in the lower right-hand corners of the tables are constantly increasing during the simplex algorithm, they constitute lower bounds for the minimum. The objective function is thus not systematically decreased by the algorithm until the minimal value has been attained, but, on the contrary, it is constantly increased. The procedure described here is called the dual-solution method. It can always be used when a linear objective function of the independent variables x_k has coefficients > 0 and is to be made minimal.

A moment ago, we set the variables y_3, y_2—i.e., the variables on the upper border of the last table—equal to zero. For a geometric interpretation of the dual solution method, the corresponding variables on the upper border of the previous tables shall also be set equal to zero. In this way, the chain of sides A, B, C in Fig. 3 will be found, which are now outside of the convex polygon and reach the polygon only in their end point.

To understand the duality phenomenon better, we consider two linear programs which have exactly the same scheme; they shall differ only in that this scheme, in one case, is written in the normal way and, in the other case, is transposed. For instance,

	x_1	x_2	x_3	1
$y_1 =$	a_{11}	a_{12}	a_{13}	c_1
$y_2 =$	a_{21}	a_{22}	a_{23}	c_2
$z =$	a_1	a_2	a_3	c

and

	$v_1 =$	$v_2 =$	$v_3 =$	$w =$
$-u_1$	a_{11}	a_{12}	a_{13}	c_1
$-u_2$	a_{21}	a_{22}	a_{23}	c_2
1	a_1	a_2	a_3	c

(63)

In the upper left program, the minimum is to be found; i.e.,

$$z = a_1 x_1 + a_2 x_2 + a_3 x_3 + c = \min$$

subject to the constraints

$$y_1 = a_{11} x_1 + a_{12} x_2 + a_{13} x_3 + c_1 \geq 0,$$

$$y_2 = a_{21} x_1 + a_{22} x_2 + a_{23} x_3 + c_2 \geq 0$$

$$x_1 \geq 0, \qquad x_2 \geq 0, \qquad x_3 \geq 0.$$

In the lower right program, the maximum is to be found; i.e.,

$$w = -c_1 u_1 - c_2 u_2 + c = \max$$

subject to the constraints

$$v_1 = -a_{11} u_1 - a_{21} u_2 + a_1 \geq 0, \qquad v_2 = -a_{12} u_1 - a_{22} u_2 + a_2 \geq 0$$

$$v_3 = -a_{13} u_1 - a_{23} u_2 + a_3 \geq 0, \qquad u_1 \geq 0, \qquad u_2 \geq 0.$$

Two such programs are called dual to each other. Assuming that the last row is positive,

$$a_1 > 0, \qquad a_2 > 0, \qquad a_3 > 0,$$

they can be solved simultaneously by applying the simplex algorithm to the joint scheme. For the upper left program this is the dual solution method, but for the lower right program it is the original method of 2.2. The value appearing in the lower right-hand corner of the last table is the desired minimum of the upper left program and, at the same time, the desired maximum for the lower right program.

This important result is formulated in brief as follows:

Duality Theorem of Linear Programming. *Two dual programs have the same extremum value.*

2.6 Application to Game Theory

Two people with the names of "Max" and "Min" are playing a game called "Morra." This game consists in having both participants stretch out a number of fingers of their right hands at the same time. For simplicity's sake, we shall allow Max to stretch out either one, two, or three fingers, while Min will be limited to extending only two fingers at the most. As for the rest, the rules of the game are given by the following table:

Max

		(1)	(2)	(3)
Min	(1)	2	6	1
	(2)	4	3	3

(64)

This table is meant as follows: When Max, for example, extends two fingers (column 2) and Min one (row 1), Min has to pay Max the sum of 6 dollars. We shall assume that this step is repeated many times, i.e., that Max and Min extend a permitted number of fingers in rapid succession—but always simultaneously. Matrix (64) indicates the pay-off to be made by Min after each step of the game; it is therefore called the pay-off matrix. For the moment, we shall assume that all its elements are > 0, which is, of course, grossly unfair to Min, since Max is constantly winning. In spite of this flaw which will be remedied later, we can ask for the optimum overall method of playing the game which these players should use in order to guard their interests as much as possible in this conflicting situation.

To put these questions more specifically, we will first identify ourselves with Max. Assume that this player decides to select his possibilities (1), (2), (3) in (64)—also called strategies—at random, yet with certain probabilities p_1, p_2, p_3. These probabilities are positive numbers with the sum $= 1$:

$$p_1 \geq 0, \qquad p_2 \geq 0, \qquad p_3 \geq 0 \tag{65}$$

$$p_1 + p_2 + p_3 = 1. \tag{66}$$

If, for example, $p_1 = 0.25$; $p_2 = 0.75$; $p_3 = 0$, then this means that Max shall choose his strategy (2) at an average of three times as often as (1), whereas strategy (3) is not to be used at all. Max, of course, intends to make the payment maximal (that is why he was called Max in the first place!) by means of appropriate selection of p_1, p_2, p_3. In order to bring this about, he goes through the following considerations. If Min were to play his strategy (1) constantly (i.e., always extending one finger only), then, according to (64), the average payment in a long succession of games would be

$$P_1 = 2p_1 + 6p_2 + p_3.$$

Similarly, if Min would constantly play his strategy (2),

$$P_2 = 4p_1 + 3p_2 + 3p_3.$$

The quantities P_1 and P_2 are certainly > 0 because of (65), (66) and the assumption concerning the matrix. Let $P > 0$ be a lower bound of P_1, P_2:

$$2p_1 + 6p_2 + p_3 \geq P, \qquad 4p_1 + 3p_2 + 3p_3 \geq P. \tag{67}$$

Of course, Min is now also playing his strategies with certain probabilities q_1, q_2, whereby

$$q_1 \geq 0, \qquad q_2 \geq 0, \qquad q_1 + q_2 = 1. \tag{68}$$

Therefore, the average payment is actually

$$q_1 P_1 + q_2 P_2 \geq (q_1 + q_2)P = P,$$

since $P_1 \geq P$, $P_2 \geq P$. In other words, P is a lower bound for the average payment, irrespective of how Min plays; i.e., with which weights q_1, q_2 he mixes his strategy. Max therefore decides to set up his game in such a way that P becomes maximal. In order to reduce this problem to a linear program, we use the independent variables

$$x_1 = \frac{p_1}{P}, \qquad x_2 = \frac{p_2}{P}, \qquad x_3 = \frac{p_3}{P}. \tag{69}$$

Since $P > 0$, all inequalities can be divided by P. From (65) it then follows that

$$x_1 \geq 0, \qquad x_2 \geq 0, \qquad x_3 \geq 0 \tag{70}$$

and from (67) that

$$\begin{aligned}
y_1 &= 2x_1 + 6x_2 + x_3 \ - 1 \geq 0, \\
y_2 &= 4x_1 + 3x_2 + 3x_3 - 1 \geq 0.
\end{aligned} \tag{71}$$

Finally, condition (66) yields

$$z = \frac{1}{P} = x_1 + x_2 + x_3. \tag{72}$$

Since P is to become maximal, the reciprocal value z has to become minimal. Inasmuch as we are now concerned with a minimum problem, we shall use the dual method of 2.5 for the solution of the linear program (70)–(72) and write the scheme in the normal form

	x_1	x_2	x_3	1
$y_1 =$	2	6	1	-1
$y_2 =$	4	3	3	-1
$z =$	1	1	1	0

$$\tag{73}$$

From this we derive the

Rule. *A two-person game is changed into a linear program by adding a border to the pay-off matrix consisting of a row of elements $(+1)$ and a column of elements (-1).*

Since the last row of the scheme is positive, the simplex algorithm can begin right away; after some calculation with fractions, we obtain:

$$x_1 = \frac{1}{6}, \qquad x_2 = \frac{1}{9}, \qquad x_3 = 0, \qquad z = \frac{5}{18}.$$

(72) then yields the maximal value of P,

$$P_{\max} = \frac{18}{5} = 3.6; \tag{74}$$

and from (69) we therefore obtain the optimal probabilities

$$\underline{\underline{p_1 = 0.6}}, \qquad \underline{\underline{p_2 = 0.4}}, \qquad \underline{\underline{p_3 = 0}}.$$

Hence, Max should never extend three fingers; this strategy remains inactive. He has to use his strategies (1) and (2) with frequencies which have a ratio of $3:2$. However, the main result (74) affirms that Max can enforce an average payment of at least 3.6.

We still have to present similar considerations for Min, whose probabilities were already denoted above by q_1, q_2. If Max constantly uses strategy (1) or (2), the columns in (64) then, respectively, yield the average payments:

$$Q_1 = 2q_1 + 4q_2, \qquad Q_2 = 6q_1 + 3q_2, \qquad Q_3 = q_1 + 3q_2.$$

Let Q be an upper bound for these values:

$$2q_1 + 4q_2 \leq Q, \qquad 6q_1 + 3q_2 \leq Q, \qquad q_1 + 3q_2 \leq Q.$$

Min understandably wants to achieve as low a payment as possible; he therefore will want to make Q minimal. Using the variables

$$u_1 = \frac{q_1}{Q}, \qquad u_2 = \frac{q_2}{Q}, \qquad w = \frac{1}{Q}$$

and considering (68), we obtain the following linear program:

$$u_1 \geq 0, \qquad u_2 \geq 0$$

$$v_1 = -2u_1 - 4u_2 + 1 \geq 0, \qquad v_2 = -6u_1 - 3u_2 + 1 \geq 0$$

$$v_3 = -u_1 - 3u_2 + 1 \geq 0$$

$$w = u_1 + u_2 = \max.$$

In transposed notation the table for this reads:

	$v_1 =$	$v_2 =$	$v_3 =$	$w =$
$-u_1$	2	6	1	-1
$-u_2$	4	3	3	-1
1	1	1	1	0

(75)

However, according to the terminology at the end of 2.5, this is precisely the dual program to the program (73) of our player Max. Then, according to the duality theorem of linear programming, both programs have the same extremum value; the desired minimal value of Q therefore is

$$\underline{\underline{Q_{\min} = P_{\max} = 3.6}}$$

Hence, Min can enforce the average payment to reach at most 3.6. As we have seen, Max can enforce at least this payment. Moreover, it now follows that, if his opponent plays skillfully, he will not be able to get a larger amount. In short, the conflicting situation resolves itself in a compromise. The common value 3.6 of $Q_{\min}$ and $P_{\max}$ is called the value of the game.

As a result of the simplex algorithm applied to the tables (73) and (75), (as already done by Max), we have

$$u_1 = \frac{1}{18}, \qquad u_2 = \frac{4}{18}, \qquad \underline{\underline{q_1 = 0.2}}, \qquad \underline{\underline{q_2 = 0.8.}}$$

The relation $Q_{\min} = P_{\max}$ (obviously valid for every pay-off matrix) is called the *duality theorem of game theory*.

A game is not essentially altered by addition of a constant a to all the elements of the pay-off matrix. In fact, if we do this in our example, we obtain for Max, instead of the payments P_1, P_2, the payments

$$(2 + a)p_1 + (6 + a)p_2 + (1 + a)p_3$$
$$= 2p_1 + 6p_2 + p_3 + a(p_1 + p_2 + p_3) = P_1 + a$$

and similarly $(P_2 + a)$. Now the quantity $(P + a)$ constitutes a lower bound for both numbers and Max will therefore make $(P + a)$ maximal. Instead of this, he could just as well make P maximal. Consequently everything remains the same and only the value of the game is increased by a. This observation removes the restrictive assumption that all the matrix elements are to be positive.

If we therefore subtract the value 3.6 of our game from all the elements of the matrix (64), we obtain a game with the value zero, which is fair to both players. But, at the same time, it ceases to be a game since both players now know that, in the long run, they can neither win nor lose. Some good examples of two-person games are given in [6].

2.61 Games with Saddle Point

Let us consider the following game with a pay-off matrix which has only elements > 0, and which we immediately complete to a linear program:

	x_1	x_2	x_3	x_4	1
$y_1 =$	a_{11}	a_{12}	a_{13}	a_{14}	-1
$y_2 =$	a_{21}	a_{22}	a_{23}	a_{24}	-1
$y_3 =$	a_{31}	a_{32}	a_{33}	a_{34}	-1
$z =$	1	1	1	1	0

Now assume that there is an element (e.g., a_{23}) called a saddle element, which satisfies the following conditions:

1. It is $\geq$ any other element in its row.
2. It is $\leq$ any other element in its column.

We choose the row of the saddle element as the pivot row for the first exchange-step of the simplex procedure. The characteristic quotients are

$$\frac{1}{a_{21}}, \quad \frac{1}{a_{22}}, \quad \frac{1}{a_{23}}, \quad \frac{1}{a_{24}}.$$

Because of Condition 1, the third quotient (belonging to the saddle element) is here the smallest one; according to Rule 2 of the simplex algorithm (see 2.2), it then follows that the saddle also becomes the pivot. Following the exchange, the values

$$-1 + \frac{a_{13}}{a_{23}}, \quad \frac{1}{a_{23}}, \quad -1 + \frac{a_{33}}{a_{23}}$$

appear in the column of the constants. Due to Condition 2, these values are ≥ 0. As a result, the simplex algorithm stops after this first step.

On the upper border of the new table we find the variables

$$x_1, \quad x_2, \quad y_2, \quad x_4$$

which are set equal to zero. Now, according to (69), the x_i are proportional to the probabilities p_i of the strategies of Max; from this follows that

$$p_1 = 0, \quad p_2 = 0, \quad p_4 = 0$$

and finally also that $p_3 = 1$ (since the sum of the probabilities must be $= 1$). In other words, Max must always use that strategy which is given by the saddle column; all other strategies are inactive. In the same way, Min always has to play the saddle row. Incidentally, the value of the game is equal to the saddle element.

2.7 Chebyshev Approximation

A surveyor wants to measure the length of two adjoining segments AB and BC of a straight line somewhere in a tract of land. Measuring each segment separately with a tape, he finds that $AB = 15$ yards, $BC = 6$ yards. As a control, he also measures the entire line and finds that $AC = 20$ yards. These three measurements evidently contradict each other and this will always be so—due to unavoidable errors in measurement—whenever excess measurements are made to determine unknown quantities. In reality, the contradiction will, of course, not have the order of magnitude of yards as here, but rather of inches or less. The problem then arises of finding values x_1, x_2 for the two segments which will fit the three measurements best, i.e., which minimize the contradiction. While the measurements state that

$$x_1 - 15 = 0, \qquad x_2 - 6 = 0, \qquad x_1 + x_2 - 20 = 0$$

the equations for the unknowns x_1, x_2 may not be written in this way. The reason is that if one were to substitute any values for x_1, x_2, certainly not all right-hand sides would be equal to zero but would assume values r_1, r_2, r_3, which are called *residual* values. We therefore have to formulate the equations as follows:

$$x_1 - 15 = r_1, \qquad x_2 - 6 = r_2, \qquad x_1 + x_2 - 20 = r_3. \tag{76}$$

They are called the *error equations* of the given approximation problem. Our own problem now obviously is to make the absolute values of these residuals small. Before we go ahead with this, we shall put the problem a little more colorfully. Assume that the surveyor also determines the length of the entire line by pacing it and that he finds $AC = 22$ yards. At first we will then be tempted to write the additional error equation as:

$$x_1 + x_2 - 22 = r_4. \tag{77}$$

However, measuring by paces is certainly less accurate than measuring with a tape—let us suppose that it is 10 times as inaccurate. To take this into account, we multiply the left side of (77) with the reciprocal value 0.1, and thus write:

$$0.1x_1 + 0.1x_2 - 2.2 = r_4$$

with another r_4. This has the result that the absolute terms 15, 6, 20, 2.2

in the four error equations have the same degree of accuracy and that therefore the four error equations have the same weight in the determination of
the unknowns.

Altogether, we obtain the approximation problem

$$
\begin{aligned}
x_1 \qquad\qquad - 15 &= r_1 \\
x_2 - 6 &= r_2 \\
x_1 + \qquad x_2 - 20 &= r_3 \\
0.1x_1 + 0.1x_2 - 2.2 &= r_4
\end{aligned}
\tag{78}
$$

With many approximation problems, we are not completely in the dark
about the values of the unknowns but can give at least approximate values
for them.   In the example, the measured values 15 and 6 are certainly good
approximate values for the separate segments.   In line with the *correction
principle* (1.33), we therefore introduce the corrections

$$
\Delta x_1 = x_1 - 15, \qquad \Delta x_2 = x_2 - 6 \tag{79}
$$

as new unknowns. This results in the following definitive form of the
approximation problem:

$$
\begin{aligned}
\Delta x_1 \qquad\qquad\qquad &= r_1 \\
\Delta x_2 \qquad\qquad &= r_2 \\
\Delta x_1 + \qquad \Delta x_2 + 1 &= r_3 \\
0.1\,\Delta x_1 + 0.1\,\Delta x_2 - 0.1 &= r_4
\end{aligned}
\tag{80}
$$

In order now to make the residuals small let r be an upper bound of their
absolute values:

$$
|r_1| \le r, \qquad |r_2| \le r, \qquad |r_3| \le r, \qquad |r_4| \le r. \tag{81}
$$

The *Chebyshev-approximation principle* then consists of minimizing the
quantity r through appropriate selection of the unknowns.   Hence, in view
of (81), all residual values will be made at least as small as this minimal
value.   To realize this principle through a linear program we first note that,
for example, the condition $|r_1| \le r$ means that the quantity r_1 lies between
$(+r)$ and $(-r)$.   This condition can therefore be replaced by the two
inequalities

$$
r_1 \le +r \qquad \text{and} \qquad r_1 \ge -r.
$$

Together with (80) this yields 8 inequalities:

$$
\left.
\begin{array}{l}
\left|
\begin{array}{l}
r - r_1 = -\quad \Delta x_1 \qquad\qquad\qquad\; + r \geq 0 \\
r - r_2 = \qquad\quad\; -\quad \Delta x_2 \qquad\quad + r \geq 0 \\
r - r_3 = -\quad \Delta x_1 -\quad \Delta x_2 - 1 \;\; + r \geq 0 \\
r - r_4 = -0.1\Delta x_1 - 0.1\Delta x_2 + 0.1 + r \geq 0
\end{array}
\right| \\[6pt]
\hline
\left|
\begin{array}{l}
r + r_1 = \qquad \Delta x_1 \qquad\qquad\qquad\; + r \geq 0 \\
r + r_2 = \qquad\qquad\quad \Delta x_2 \qquad\quad + r \geq 0 \\
r + r_3 = \qquad \Delta x_1 + \quad \Delta x_2 + 1 \;\; + r \geq 0 \\
r + r_4 = \quad 0.1\Delta x_1 + 0.1\Delta x_2 - 0.1 + r \geq 0
\end{array}
\right|
\end{array}
\right\} \qquad (82)
$$

as well as the requirement $r = \min$.

In this way we obtained a linear program with the independent variables Δx_1, Δx_2, r, where the objective function is simply equal to the last variable r. Even though we could apply our methods directly to this program, it is more convenient to introduce the new variables

$$
\xi_i = \frac{\Delta x_1}{r}, \qquad \xi_2 = \frac{\Delta x_2}{r}, \qquad \xi_3 = \frac{1}{r} \qquad (83)
$$

and to divide all inequalities by r (as upper bound of a set of absolute values r is certainly positive). Then the following program results:

$$
\left.
\begin{array}{l}
\left|
\begin{array}{l}
y_1 = 1 - \dfrac{r_1}{r} = -\quad \xi_1 \qquad\qquad\qquad + 1 \geq 0 \\[6pt]
y_2 = 1 - \dfrac{r_2}{r} = \qquad\quad -\quad \xi_2 \qquad\quad + 1 \geq 0 \\[6pt]
y_3 = 1 - \dfrac{r_3}{r} = -\quad \xi_1 -\quad \xi_2 -\quad \xi_3 + 1 \geq 0 \\[6pt]
y_4 = 1 - \dfrac{r_4}{r} = -0.1\xi_1 - 0.1\xi_2 - 0.1\xi_3 + 1 \geq 0
\end{array}
\right| \\[6pt]
\hline
\left|
\begin{array}{l}
y_1' = 1 + \dfrac{r_1}{r} = \quad \xi_1 \qquad\qquad\qquad + 1 \geq 0 \\[6pt]
y_2' = 1 + \dfrac{r_2}{r} = \qquad\quad \xi_2 \qquad\qquad + 1 \geq 0 \\[6pt]
y_3' = 1 + \dfrac{r_3}{r} = \quad \xi_1 + \quad \xi_2 + \quad \xi_3 + 1 \geq 0 \\[6pt]
y_4' = 1 + \dfrac{r_4}{r} = \quad 0.1\xi_1 + 0.1\xi_2 + 0.1\xi_3 + 1 \geq 0
\end{array}
\right|
\end{array}
\right\} \qquad (84)
$$

$z = \xi_3$ has to be made maximal. Hence we write the table in transposed form:

	$y_1 =$	$y_2 =$	$y_3 =$	$y_4 =$	$y_1' =$	$y_2' =$	$y_3' =$	$y_4' =$	$z =$
$-\xi_1$	1	0	1	0.1	-1	0	-1	-0.1	0
$-\xi_2$	0	1	1	0.1	0	-1	-1	-0.1	0
$-\xi_3$	0	0	1	-0.1	0	0	-1	0.1	-1
1	1	1	1	1	1	1	1	1	0

The given error equations (80) appear in the y-columns while the y'-columns are obtained from the obvious relation

$$\boxed{y_i + y_i' = 2} \tag{85}$$

This same relation can also be applied in the following to simplify the calculations. The last row of the program is fortunately positive. First of all, we will have to eliminate the free variables ξ_1, ξ_2, ξ_3 according to the methods set forth in 2.4. We begin with the elimination of ξ_3 and find the table

	$y_1 =$	$y_2 =$	$y_4 =$	$y_1' =$	$y_2' =$	$y_3' =$	$y_4' =$	$z =$	
$-\xi_1$	1	0	0.2	-1	0	0	-0.2	1	
$-\xi_2$	0	1	0.2	0	-1	0	-0.2	1	
$-y_3$	0	0	0.1	0	0	1	-0.1	1	(86)
1	1	1	1.1	1	1	2	0.9	1	

In order to discuss this result, we set the variables on the left-hand border equal to zero, as in 2.3. [Since the last row is positive, all constraints (84) and hence also those in (81) are satisfied.] According to (83), $\xi_1 = \xi_2 = 0$ implies that also

$$\Delta x_1 = \Delta x_2 = 0,$$

and so we find ourselves at the approximation point selected above, namely, $x_1 = 15$, $x_2 = 6$. Because of $y_3 = 0$ and (84), we have for its residual $r_3 = r$, and, because of (83), this is the reciprocal value of $z = \xi_3$ which appears in the lower right-hand corner of the table.

According to (81), the remaining residuals are at most equal to r.   Hence the value of r is the maximum of the absolute values of the residuals at the approximation point.

The column which had been left out at this EX-step furthermore yields

$$\xi_3 = -\xi_1 - \xi_2 - y_3 + 1. \tag{87}$$

Elimination of ξ_1 now results in

	$y_1 =$	$y_2 =$	$y_4 =$	$y_2' =$	$y_3' =$	$y_4' =$	$z =$
$-y_1'$	1	0	0.2	0	0	-0.2	1
$-\xi_2$	0	1	0.2	-1	0	-0.2	1
$-y_3$	0	0	0.1	0	1	-0.1	1
1	2	1	1.3	1	2	0.7	2

and

$$\xi_1 = y_1' - 1. \tag{88}$$

Again we set the variables on the left-hand border equal to zero; because of (88) this yields $\xi_1 = -1$, $\xi_2 = 0$.   From this, together with (83), certain corrections follow, as well as values of x_1, x_2, but we shall not compute them. In the corresponding point (x_1, x_2) the maximum of the absolute values of the residuals is again the reciprocal value of the element in the lower right-hand corner.   However, the considerations in 2.4 guarantee that the element in the lower right-hand corner increases in the transition from (86) to (88).   Hence, in the transition from the approximation point to the new point (x_1, x_2), the *maximum absolute value of the residuals* has certainly *decreased,* and this is the case in all the following EX-steps.   The procedure therefore consists in decreasing methodically and successively the largest residual.

Finally, the elimination of ξ_2 leads to

	$y_1 =$	$y_2 =$	$y_4 =$	$y_3' =$	$y_4' =$	$z =$
$-y_1'$	1	0	0.2	0	-0.2	1
$-y_2'$	0	1	0.2	0	-0.2	1
$-y_3$	0	0	0.1	1	-0.1	1
1	2	2	1.5	2	0.5	3

and

$$\xi_2 = y_2' - 1. \tag{89}$$

Now the simplex algorithm should actually begin. By coincidence, however, the last column is already positive, and we are therefore finished. Setting the variables on the left border equal to zero, we find according to (84) that

$$r_1 = -r, \qquad r_2 = -r, \qquad r_3 = +r.$$

Upon completion of the approximation process, three residuals must therefore be equal in absolute value and, because of (81), the fourth is at most that large. From (89) and (88) it further follows that

$$\xi_2 = -1, \qquad \xi_1 = -1$$

and the lower right-hand corner of the table yields $r = \frac{1}{3}$. Finally, we obtain from the transformation equations (83) and (79) the following end result of the approximation problem:

$$x_1 = 14\tfrac{2}{3}, \qquad x_2 = 5\tfrac{2}{3}.$$

At this point, the maximum of the residual values equals $r = \frac{1}{3}$. A point where the absolute values of the residuals have a smaller maximum is not possible, since, otherwise, an upper bound for the residual values could be found which is smaller than $\frac{1}{3}$. But, according to our construction, $\frac{1}{3}$ is the smallest value which such a bound can have. Consequently, the *Chebyshev approximation principle* is equivalent to minimizing the *maximal absolute value of the residuals*.

It should further be pointed out that with the help of the relations (85) the computational technique can easily be refined in such a way that only the functions y_i have to be carried in the tables, and not the functions y_i' as well.

The general approximation problem of this type has the following form. Consider m error equations in n unknowns

$$\sum_{k=1}^{n} a_{ik}x_k + c_i = r_i, \qquad i = 1, 2, \dots, m$$

whereby the c_i measurements are of equal accuracy. m must be $> n$; otherwise there are no excess measurements at all. Hence, in the table of the corresponding linear program as well as in all tables resulting from it in the course of the simplex procedure, there are $(n + 1)$ variables on the left border instead of three variables. Accordingly, at the end of the procedure, $(n + 1)$ variables have to be set equal to zero. As in the example, this leads to the conclusion that after the approximation procedure $(n + 1)$ *residuals must have the same absolute value while the absolute values of the remaining residuals can be at most as large.* Although this criterion is necessary for the completion of a Chebyshev approximation, it is not sufficient. In fact, it is also satisfied in every intermediate stage of the simplex algorithm.)

3 Least-Squares Approximation and Definite Problems

We have seen that contradictions are always the result of unavoidable measurement errors when excess measurements have been made for the determination of unknown quantities. To minimize these contradictions, the Chebyshev principle which we just described is not used in most cases, but rather the Gauss method of least-squares. There are two decisive reasons for this. First of all, from the point of computational technique, the Gauss method is easier to handle, and secondly, in case the measurement errors are randomly distributed according to the laws of probability, only the latter method is really appropriate for the approximation problem. The Chebyshev principle retains its importance, however, in another type of approximation problem to be discussed later in the book. A reader who is interested only in Gauss approximations should read 2.7 as an introduction, at least up to equation (80).

3.1 The Method of Least-Squares

Let the following m error equations for the determination of the two unknowns x_1, x_2 be given:

$$
\left.
\begin{aligned}
a_{11}x_1 + a_{12}x_2 + c_1 &= r_1 \\
a_{21}x_2 + a_{22}x_2 + c_2 &= r_2 \\
\cdot \quad \cdot \quad \cdot \quad \cdot \quad \cdot \quad \cdot \quad \cdot \quad \cdot \quad \cdot & \\
a_{m1}x_1 + a_{m2}x_2 + c_m &= r_m
\end{aligned}
\right\}
\tag{90}
$$

The c_i are measurements of equal accuracy while the r_i are the residuals. In order to give a geometric justification of the Gauss principle, we assume for a moment that $m = 3$, i.e., that there is only one excess equation. The three residuals, r_1, r_2, r_3, then can be interpreted as the coordinates of a point R in a three-dimensional Cartesian coordinate system (Fig. 4). The closer R is to the origin O of the coordinate system, the smaller the residuals will be. For this reason we want to minimize the distance

$$OR = \sqrt{r_1^2 + r_2^2 + r_3^2}$$

through appropriate selection of x_1, x_2. Alternately we can minimize the square of this distance

$$r_1^2 + r_2^2 + r_3^2.$$

In the more general case (90) of an arbitrary number of error equations, we use as approximation principle

$$r_1^2 + r_2^2 + \ldots + r_m^2 = \min. \tag{91}$$

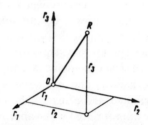

FIG. 4.   Least-squares principle.

In order to compute this quantity, we employ the concept of the "scalar product," already mentioned briefly in 1.34.

Consider two rows of numbers[1]

$$p_1, p_2, p_3, \ldots, p_m$$

and

$$q_1, q_2, q_3, \ldots, q_m.$$

Their scalar product is defined as the quantity

$$(p, q) = p_1 q_1 + p_2 q_2 + p_3 q_3 + \ldots + p_m q_m.$$

Evidently, the commutative law

$$(p, q) = (q, p)$$

holds. In particular, the product of the first row of numbers with itself is

$$(p, p) = p_1^2 + p_2^2 + p_3^2 + \ldots + p_m^2.$$

With this convenient notation, (91) can be written as

$$(r, r) = \min. \tag{92}$$

[1] A row of numbers $p_1, p_2, \ldots, p_m$ is also called a *vector*.

From (90) it now follows that

$$(r,r) = (a_1, a_1)x_1^2 + 2(a_1, a_2)x_1x_2 + (a_2, a_2)x_2^2$$
$$+ 2(a_1, c)x_1 + 2(a_2, c)x_2 + (c, c). \tag{93}$$

Here, for example, (a_1, a_2) is the scalar product of the first two columns of (90); hence,

$$(a_1, a_2) = a_{11}a_{12} + a_{21}a_{22} + \ldots + a_{m1}a_{m2}$$

and (a_1, a_1) is the scalar product of the first column with itself:

$$(a_1, a_1) = a_{11}^2 + a_{21}^2 + \ldots + a_{m1}^2.$$

According to (93), the quantity (r, r) to be minimized is a *quadratic function* of the independent variables x_1, x_2. It is composed of a homogeneous quadratic function in the first row of (93) and a linear function in the second row. A homogeneous quadratic function is also called a *quadratic form*.

In order to make (r, r) minimal, (93) will first have to be differentiated with respect to x_1 and then with respect to x_2. This yields the two linear equations

$$(a_1, a_1)x_1 + (a_1, a_2)x_2 + (a_1, c) = 0$$
$$(a_2, a_1)x_1 + (a_2, a_2)x_2 + (a_2, c) = 0 \tag{94}$$

for the determination of the unknowns x_1, x_2. They are called the *Gauss normal equations*. It must be observed that in the corresponding matrix

$$\begin{vmatrix} (a_1, a_1) & (a_1, a_2) \\ (a_2, a_1) & (a_2, a_2) \end{vmatrix}$$

both elements (a_1, a_2) and (a_2, a_1)—located symmetrically with respect to the diagonal—are equal. We therefore speak of a *symmetric matrix*. As first rule for the formation of the normal equations, one should remember that *the coefficients in the normal equations are scalar products of the coefficient columns of the error equations.*[1]

---

[1] If only two equations (90) with two unknowns are given, i.e., if $m = 2$, then the solutions of the normal equations are identical with the solutions of the equations

$$a_{11}x_1 + a_{12}x_2 + c_1 = 0$$
$$a_{21}x_1 + a_{22}x_2 + c_2 = 0.$$

In fact, the solutions of the latter equations imply that the r_1, r_2 in (90) are equal to zero; hence, they certainly reduce $(r, r) = r_1^2 + r_2^2$ to the smallest possible value, namely, zero. The analogous notation applies to n equations with n unknowns.

Row sums can be used to check the correctness of the normal equations. Let

$$s_i = a_{i1} + a_{i2} + c_i$$

be the row sum of the ith error equation. Then we can easily ascertain through calculation that for the row sums in both normal equations

$$(a_1, a_1) + (a_1, a_2) + (a_1, c) = (a_1, s)$$

$$(a_2, a_1) + (a_2, a_2) + (a_2, c) = (a_2, s).$$

Here s is the column of the s_i.

3.11   The Mean Residual

After solving the normal equations (94), we can substitute the resulting values of the unknowns in the error equations (90). This yields the *residuals after the approximation* which we shall call $\rho_1, \rho_2, \ldots, \rho_m$.

For any pair of values x_1, x_2 and the corresponding residuals $r_1, r_2, \ldots, r_m$ we certainly have

$$(\rho, \rho) = \rho_1^2 + \rho_2^2 + \cdots + \rho_m^2 \leq r_1^2 + r_2^2 + \cdots + r_m^2. \tag{95}$$

This follows from the fact that the left-hand side is, after all, the minimum value attainable by the sum of the squares of the residuals.

Now let $r_{\max}$ be the largest of the numbers

$$|r_1|, |r_2|, \ldots, |r_m|,$$

so that we have

$$r_1^2 + r_2^2 + \cdots + r_m^2 \leq m r_{\max}^2. \tag{96}$$

The two inequalities (95), (96) then imply that

$$(\rho, \rho) \leq m r_{\max}^2.$$

Accordingly, if we introduce the *mean residual*[1] *after the approximation*

$$\bar{\rho} = \sqrt{\frac{(\rho, \rho)}{m}}, \tag{97}$$

it follows that

$$r_{\max} \geq \bar{\rho}. \tag{98}$$

Therefore: *For arbitrary values of the unknowns, the maximal absolute value of the residuals is at least as large as the mean residual after the approximation.* This bound ρ can not be lowered by any other approximation procedure.

This theorem establishes in particular a relationship with the Chebyshev

---

[1] This concept is not to be confused with the concept of the average error, used in calculus of observations. Its definition is founded on probability-theoretical reasons.

approximation which makes r_{max} as small as possible and for which we therefore have

$$r_{max} \leq \rho_{max} \qquad (99)$$

Here ρ_{max} denotes the largest among the residual values

$$|\rho_1|, |\rho_2|, \ldots, |\rho_m|$$

of the Gauss approximation. In conclusion, (98) and (99) give us the following inequality for the maximum residual value after the Chebyshev approximation:

$$\boxed{\bar{\rho} \leq r_{max} \leq \rho_{max}} \qquad (100)$$

This result is useful in deciding whether it is worth it to improve a Gauss approximation by changing it into a Chebyshev approximation. If this is desirable, the results x_1, x_2 of the Gauss approximation can be taken as initial values for the Chebyshev approximation.

1. **Example.** The error equations (80)

$$\begin{aligned}
\Delta x_1 &&&= r_1 \\
&& \Delta x_2 && = r_2 \\
\Delta x_1 +&& \Delta x_2 + 1 &= r_3 \\
0.1\Delta x_1 +& 0.1\Delta x_2 - 0.1 &= r_4
\end{aligned}$$

are to be *smoothed* by the method of least squares.
Normal equations:

$$\begin{aligned}
2.01\Delta x_1 + 1.01\Delta x_2 + 0.99 &= 0 \\
1.01\Delta x_1 + 2.01\Delta x_2 + 0.99 &= 0
\end{aligned}$$

Solutions:

$$\Delta x_1 + \underline{\underline{-0.3278}}, \qquad \Delta x_2 + \underline{\underline{-0.3278}}.$$

Residuals following the approximation:

$$-0.3278, \qquad -0.3278, \qquad 0.3444, \qquad -0.1656.$$

Mean residual:

$$\bar{\rho} = \underline{\underline{0.3004}}.$$

For the Chebyshev approximation, therefore:

$$0.3004 \leq r_{max} \leq 0.3444.$$

In 2.7 we found

$$r_{\max} = 0.3333.$$

When formulating applied approximation problems, one must pay careful attention to a suitable choice of unknowns. First and foremost, one must introduce only as many unknowns as needed and no more. This means that no exact relation, i.e., no relation independent of the measurements, shall exist between the unknowns. In the more advanced development of approximation theory, such conditions are permitted (approximation with "side conditions"), but we will have to forbid them in our own elementary considerations. Second, the error equations have to be set up correctly. As far as this can be determined *a priori*, their absolute terms should have the same degree of accuracy; otherwise, this is to be achieved by multiplying the error equations with suitable coefficients (see 2.7). It is forbidden, for example, to add two error equations since the absolute term of the resulting equation will certainly be less accurate than the absolute terms of the original equations. Other algebraic manipulations with the error equations are likewise not permitted.

2. Example. The three angles of a triangle have been measured and the results of these measurements are α_1, α_2, α_3 (in degrees). We introduce the first two angles x_1, x_2 of the triangle as unknowns. The third angle cannot be taken as another unknown since the three angles must satisfy the law for the sum of the angles in a triangle. The measurement α_3 states that

$$180° - (x_1 + x_2) = \alpha_3.$$

Hence, the correct error equations are

$$\begin{aligned}
x_1 \quad\quad\quad - \alpha_1 \;\; &= r_1 \\
x_2 \quad - \alpha_2 \;\; &= r_2 \\
x_1 + x_2 + (\alpha_3 - 180) &= r_3
\end{aligned}$$

and the normal equations are:

$$\begin{aligned}
2x_1 + \;\; x_2 + (\alpha_3 - \alpha_1 - 180) &= 0 \\
x_1 + 2x_2 + (\alpha_3 - \alpha_2 - 180) &= 0
\end{aligned}$$

They have the solution

$$x_1 = \alpha_1 + \frac{\varepsilon}{3}, \quad\quad x_2 = \alpha_2 + \frac{\varepsilon}{3},$$

where

$$\varepsilon = 180° - (\alpha_1 + \alpha_2 + \alpha_3).$$

For the third angle we obtain from the law for the sum of the angles

$$180 - (x_1 + x_2) = \alpha_3 + \frac{\varepsilon}{3}.$$

In other words, the error between the measured sum of the angles and the correct value of $180°$ has to be distributed equally among all three angles.

3.12 Several Unknowns

In this case, nothing essentially new needs to be added. Consider, for example, m error equations with three unknowns x_1, x_2, x_3:

$$a_{i1}x_1 + a_{i2}x_2 + a_{i3}x_3 + c_i = r_i, \qquad i = 1, 2, \ldots, m. \tag{101}$$

We obtain for the sum of the squares of the residuals

$$
\begin{aligned}
(r, r) = (a_1, a_1)x_1^2 &+ 2(a_1, a_2)x_1x_2 + 2(a_1, a_3)x_1x_3 \\
&+ \ (a_2, a_2)x_2^2 \ \ + 2(a_2, a_3)x_2x_3 \\
&\qquad\qquad\quad + \ (a_3, a_3)x_3^2 \\
+ 2(a_1, c)x_1 + 2(a_2, c)x_2 \ &+ 2(a_3, c)x_3 + (c, c).
\end{aligned}
\tag{102}
$$

Here, for instance, (a_2, a_3) is the scalar product of the second and third column of coefficients in the error equations and (a_1, c) is the scalar product of the first column of coefficients and the column of constants. Again, the function (r, r) is composed of a quadratic form in the first three rows of (102) and a linear function in the fourth row. The normal equations have the form

$$
\left.
\begin{aligned}
(a_1, a_1)x_1 + (a_1, a_2)x_2 + (a_1, a_3)x_3 + (a_1, c) = 0 \\
(a_2, a_1)x_1 + (a_2, a_2)x_2 + (a_2, a_3)x_3 + (a_2, c) = 0 \\
(a_3, a_1)x_1 + (a_3, a_2)x_2 + (a_3, a_3)x_3 + (a_3, c) = 0
\end{aligned}
\right\}
\tag{103}
$$

Their matrix is again symmetric, i.e., two elements in positions symmetric to the diagonal are equal. Figure 5 illustrates this; in this figure, equal elements

FIG. 5.   Symmetric matrix.

appear at the ends of an arrow. Equation (97) for the mean residual remains the same, and so does the inequality (100).

3. Example. While surveying a terrain, it is desired to determine the altitudes x_1, x_2, x_3, x_4 of four levels. This is done, on the one hand, by measuring the altitudes directly and, on the other hand, by measuring the six altitude differences between the four levels. The measurements are given in Fig. 6. Smooth these results.

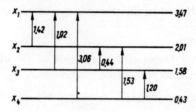

FIG. 6.   Smoothing with 4 unknowns.

Error equations:

x_1	x_2	x_3	x_4	1
1				−3.47
	1			−2.01
		1		−1.58
			1	−0.43
1	−1			−1.42
1		−1		−1.92
1			−1	−3.06
	1	−1		−0.44
	1		−1	−1.53
		1	−1	−1.20

Normal equations:

x_1	x_2	x_3	x_4	1
4	−1	−1	−1	−9.87
−1	4	−1	−1	−2.56
−1	−1	4	−1	−0.42
−1	−1	−1	4	+5.36

In Appendix I, Example 3, this system is solved with the help of the compact form of the Gauss algorithm. The corrected altitudes after the approximation are

$$x_1 = 3.472, \qquad x_2 = 2.010, \qquad x_3 = 1.582, \qquad x_4 = 0.426.$$

Corresponding to these values are the 10 residuals (in units of the third place after the decimal point):

| 2 | 0 | 2 | −4 | 42 | −30 | −14 | −12 | 54 | −44. |

The mean residual $= 28$.

4. Example. From a table of the exponential function $y = e^x$, we obtain the following values

$x =$	0	0.25	0.5	0.75	1
$y =$	1.0000	1.2840	1.6487	2.1170	2.7183
$t =$	−2	−1	0	1	2

(104)

(The third row will be explained shortly.) The problem is to find a parabola of the form

$$y = c_0 + c_1 x + c_2 x^2$$

which approximates these values as closely as possible. In order to be able to calculate with somewhat simpler numbers, we introduce the substitution

$$t = 4x - 2 \tag{105}$$

which transforms the interval $(0; 1)$ of the given x-values into the t-interval $(-2; +2)$, i.e., t runs through the values given in (104). The equation of the desired parabola assumes the form

$$y = a_0 + a_1 t + a_2 t^2.$$

The three coefficients a_0, a_1, a_2 are the unknowns of our problem. At $t = 2$, for instance, our parabola has the value

$$y(2) = a_0 + 2a_1 + 4a_2$$

and this value will approximate the given number 2.7183. The *approximation error* at this point

$$r_1 = y(2) - 2.7183 = a_0 + 2a_1 + 4a_2 - 2.7183$$

therefore has to be made small. Hence, we obtain the error equation

$$a_0 + 2a_1 + 4a_2 - 2.7183 = r_1$$

and the residual r_1 assumes the meaning of the approximation error. Altogether we obtain the five error equations

a_0	a_1	a_2	1
1	2	4	−2.7183
1	1	1	−2.1170
1	0	0	−1.6487
1	−1	1	−1.2840
1	−2	4	−1.0000

The normal equations are

a_0	a_1	a_2	1
5	0	10	−8.7680
0	10	0	−4.2696
10	0	34	−18.2742

and they have the solutions

$$a_0 = 1.6481, \qquad a_1 = 0.4270, \qquad a_2 = 0.0527. \tag{106}$$

The residuals are (in units of the fourth decimal place)

$$-54, \quad 108, \quad -6, \quad -102, \quad 49.$$

The mean residual is 52. For the maximal approximation error r_{max} after a Chebyshev approximation, we obtain the bounds

$$52 \leq r_{max} \leq 108. \tag{107}$$

From (105) and (106) it follows that the equation of the approximating parabola is

$$y = 1.0049 + 0.8648x + 0.8432x^2. \tag{108}$$

It represents the exponential function in the interval $(0; 1)$ with about the accuracy of a slide rule.

The approximation problem we have just solved has nothing to do with measurements; the function values of the exponential function in the given table are accurate up to the last decimal. Therefore, statistics and probability theory do not apply here. Consequently, the Gauss principle loses its special significance; it is desirable to follow up with a Chebyshev approximation (see 2.7) by using the values (106) as initial approximation. In this way, the maximal residual is reduced to the value

$$r_{max} = 87$$

and for the coefficients of the equation of the parabola we find the values

$$a_0 = 1.6476, \qquad a_1 = 0.4252, \qquad a_2 = 0.0529$$

which are only slightly different from those in (106).

3.2 Definite Problems

We begin with a theoretical discussion of the error equations. Since several variables only result in more complex formulas without giving any new information, we shall restrict ourselves again to the case (90) of two variables x_1, x_2. Let us introduce the linear forms

$$\left.\begin{aligned}
\eta_1 &= a_{11}x_1 + a_{12}x_2 \\
\eta_2 &= a_{21}x_1 + a_{22}x_2 \\
&\cdots\cdots\cdots\cdots \\
\eta_m &= a_{m1}x_1 + a_{m2}x_2
\end{aligned}\right\} \qquad (109)$$

belonging to the homogeneous equations, and let us assume that the homogeneous error equations

$$a_{i1}x_1 + a_{i2}x_2 = 0, \qquad i = 1, 2, \ldots, m$$

have only the trivial solution $x_1 = 0$, $x_2 = 0$. In other words, when all η become zero, it follows that all x disappear. We now form

$$(\eta, \eta) = \eta_1^2 + \eta_2^2 + \ldots \eta_m^2 = (a_1, a_1)x_1^2 + 2(a_1, a_2)x_1x_2 + (a_2, a_2)x_2^2. \quad (110)$$

This is nothing else but the quadratic form representing the first part of the function (93) that has to be minimized. For arbitrary real x_1, x_2, this form must always have a positive value since it is the sum of the squares of the η; i.e., it is the sum of positive numbers. The value of this form can equal zero only if all η, and therefore also both x_1, x_2, are equal to zero.

Consider now any quadratic form

$$f = \alpha_{11}x_1^2 + 2\alpha_{12}x_1x_2 + \alpha_{22}x_2^2 \tag{111}$$

in the independent variables x_1, x_2. α_{11}, α_{12}, α_{22} here are any given real numbers; the factor 2 for the term in the middle was inserted in order to stress the analogy with (110). This form is called *positive-definite*, or *definite* for short, if the following two conditions are satisfied:

1. For arbitrary real values of the variables, f is always ≥ 0.
2. $f = 0$ is possible only if every one of the variables equals zero.

Using this definition we can now formulate the following result:
The quadratic forms corresponding to smoothing problems are definite.

In an analogy to (93), consider an inhomogeneous quadratic function F of the variables x_1, x_2, which is composed of a definite quadratic form f and a linear function:

$$F = \alpha_{11}x_1^2 + 2\alpha_{12}x_1x_2 + \alpha_{22}x_2^2 + 2\beta_1 x_1 + 2\beta_2 x_2 + \gamma. \tag{112}$$

The problem of minimizing such a function by suitable choice of x_1, x_2 is called a *definite problem*. Differentiation with respect to x_1, x_2 immediately leads to the "normal equations"

$$\left.\begin{aligned}\alpha_{11}x_1 + \alpha_{12}x_2 + \beta_1 = 0 \\ \alpha_{21}x_1 + \alpha_{22}x_2 + \beta_2 + 0\end{aligned}\right| \tag{113}$$

For uniformity of notation, we have introduced the number $\alpha_{21} = \alpha_{12}$. The matrix of the system of equations is again symmetric. More precisely, we are speaking of *symmetric-definite equations*.

In the case of three variables x_1, x_2, x_3 the given function F has the form

$$\begin{aligned}F = \alpha_{11}x_1^2 &+ 2\alpha_{12}x_1x_2 + 2\alpha_{13}x_1x_3 \\ &+ \ \alpha_{22}x_2^2 \ + 2\alpha_{23}x_2x_3 \\ &+ \ \alpha_{33}x_3^2 \\ &+ 2\beta_1 x_1 + 2\beta_2 x_2 + 2\beta_3 x_3 + \gamma.\end{aligned} \tag{114}$$

Here, the quadratic form f which appears in the first three rows has to be definite, i.e., it must satisfy the conditions 1., 2. The corresponding definite problem requires the minimization of F and leads to the normal equations

$$\left.\begin{aligned}\alpha_{11}x_1 + \alpha_{12}x_2 + \alpha_{13}x_3 + \beta_1 = 0 \\ \alpha_{21}x_1 + \alpha_{22}x_2 + \alpha_{23}x_3 + \beta_2 = 0 \\ \alpha_{31}x_1 + \alpha_{32}x_2 + \alpha_{33}x_3 + \beta_3 = 0\end{aligned}\right| \quad \text{where} \quad \begin{aligned}\alpha_{21} &= \alpha_{12} \\ \alpha_{31} &= \alpha_{13} \\ \alpha_{32} &= \alpha_{23}.\end{aligned} \tag{115}$$

These equations are again symmetric.   *The problems of approximation by means of the method of least-squares are therefore definite problems.*   But problems of this type also occur frequently in other areas of applied mathematics.   This is always the case, for example, when we are concerned with the equilibrium of a mechanical system under the influence of conservative forces.

Example.   Two horizontally movable point-masses m_1, m_2 are connected with two fixed points A and B by springs of equal rigidity.   In the initial position (the upper part of Fig. 7), the springs are released and under no

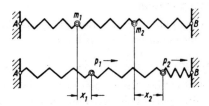

FIG. 7.   Definite problem.

tension.   Then forces p_1, p_2 are applied to the point-masses and a new equilibrium configuration results.   The problem is to determine the deflections x_1, x_2 (lower part of Fig. 7).   In the new configuration, the three individual springs possess deformation energies which have the values

$$\frac{k}{2} x_1^2, \qquad \frac{k}{2} (x_2 - x_1)^2, \qquad \frac{k}{2} x_2^2,$$

respectively.   Here k is the spring constant.   The external forces have performed the work

$$p_1 x_1 + p_2 x_2.$$

According to the fundamental variational principles of mechanics, the quadratic function

$$F = \frac{k}{2} [x_1^2 + (x_2 - x_1)^2 + x_2^2] - p_1 x_1 - p_2 x_2$$

has to become minimal.   The quadratic form in the square brackets is a sum of squares and therefore definite.   The normal equations have the form

$$
\begin{aligned}
k(\ 2x_1 - \ x_2) - p_1 &= 0 \\
k(-x_1 + 2x_2) - p_2 &= 0
\end{aligned}
\tag{116}
$$

It is very useful to accustom oneself always to formulate such *equilibrium problems* as *minimal problems of quadratic functions.* The resulting linear system of equations then is automatically symmetric, which is a great advantage. The problem of Fig. 7 had finitely many (namely two) degrees of freedom. Somewhat more complicated, but nevertheless basically of the same type, are problems of infinitely many degrees of freedom (bending of beams and plates in the theory of elasticity; stationary flows in hydrodynamics and electrodynamics, etc.). We shall return to this when we take up partial differential equations.

3.21 Geometrical Interpretation

Let us represent the quadratic function (112) graphically in an x_1, x_2-coordinate system by means of its contour lines

$$\alpha_{11}x_1^2 + 2\alpha_{12}x_1x_2 + \alpha_{22}x_2^2 + 2\beta_1 x_1 + 2\beta_2 x_2 + \gamma = \text{const.} \qquad (117)$$

Here (117) is a family of conics; these conics are ellipses since the quadratic form constituting the first part of the left-hand side of (117) is definite. We know from analytic geometry that, for an ellipse, the location of the center and of the axes, as well as the proportion of the length of these axes, does not depend on the value of the constant in the equation of the ellipse. Consequently, (117) represents a family of concentric ellipses similar to each other (Fig. 8). The value F of the quadratic function (112) increases when the

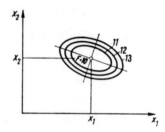

FIG. 8. Definite problem. Error ellipses.

ellipses grow bigger. If we represent these values of F as altitudes above the x_1, x_2-plane, we obtain a three-dimensional surface which has the form of a bowl—more precisely, which is an elliptic paraboloid. The lowest point of this bowl coincides with the common center of all ellipses; its coordinates are therefore the solutions of the given definite problem $F = \text{min.}$

If the function F stems from an approximation problem, i.e., if it has the form (93), then the ellipses are called *error ellipses.* For problems with three or more independent variables, the ellipses obviously are replaced by ellipsoids in a three- or higher-dimensional space.

3.3 Solution of Symmetric-Definite Equations

For the solution of the normal equations (115), the exchange method or the Gauss elimination technique is generally used (see 1.21, 1.31). In order to discuss the course of this procedure, we introduce the linear forms

$$y_1 = \alpha_{11}x_1 + \alpha_{12}x_2 + \alpha_{13}x_3$$

$$y_2 = \alpha_{21}x_1 + \alpha_{22}x_2 + \alpha_{23}x_3 \qquad (118)$$

$$y_3 = \alpha_{31}x_1 + \alpha_{32}x_2 + \alpha_{33}x_3$$

corresponding to the normal equations. In tabular form they can be written as

	x_1	x_2	x_3		
$y_1 =$	α_{11}	α_{12}	α_{13}		$\alpha_{21} = \alpha_{12}$
$y_2 =$	α_{21}	α_{22}	α_{23}	where	$\alpha_{31} = \alpha_{13}$
$y_3 =$	α_{31}	α_{32}	α_{33}		$\alpha_{32} = \alpha_{23}.$

$$(119)$$

From (118) follows that

$$x_1y_1 + x_2y_2 + x_3y_3 = \alpha_{11}x_1^2 + 2\alpha_{12}x_1x_2 + 2\alpha_{13}x_1x_3$$
$$+ \ \alpha_{22}x_2^2 + 2\alpha_{23}x_2x_3$$
$$+ \ \alpha_{33}x_3^2 \qquad (120)$$

and this is the definite quadratic form f which belongs to the given definite problem (114):

$$f = x_1y_1 + x_2y_2 + x_3y_3. \qquad (121)$$

Substitution of the special values $x_1 = 1$, $x_2 = 0$, $x_3 = 0$ yields $f = \alpha_{11}$. The first condition for definite forms (see 3.2) therefore implies that $\alpha_{11} \geq 0$. Moreover, $\alpha_{11} \neq 0$ because otherwise $f = 0$ for these special values of the variables; this, in turn, contradicts the second condition for definiteness, since not all variables are zero ($x_1 = 1$). Because $\alpha_{11} \neq 0$, we can select as pivot for a first exchange-step the element in the left-hand upper corner of the table (119). Now we shall execute the exchange in the slightly modified form which was described briefly in 1.13. In other words, we introduce the variable $(-x_1)$ on the left-hand border; for this purpose, it will be advantageous to use the following notation:

$$
\begin{array}{c|ccc}
 & y_1 & x_2 & x_3 \\
\hline
-x_1 = & -\alpha'_{11} & \alpha'_{12} & \alpha'_{13} \\
y_2 = & \alpha'_{21} & \alpha'_{22} & \alpha'_{23} \\
y_3 = & \alpha'_{31} & \alpha'_{32} & \alpha'_{33}
\end{array}
\tag{122}
$$

Again we have a symmetric matrix. For the first row and the first column, the symmetry follows from Rules 2 and 3 in 1.1. For the rest of the matrix it follows from the rectangle rule also given in 1.1. [The two rectangles belonging to symmetrically located elements (119) are mirror images of each

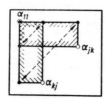

Fig. 9. Rectangle rule for a symmetric matrix: α_{11} is the pivot element; $\alpha_{jk} = \alpha_{kj}$ are the elements subjected to the calculation.

other with the diagonal acting as mirror. Hence, the numbers connected by arrows in Fig. 9 must be equal.] From (121) and (122) we obtain for our quadratic form:

$$
\begin{aligned}
f &= -(-\alpha'_{11}y_1 + \alpha'_{12}x_2 + \alpha'_{13}x_3)y_1 + (\alpha'_{21}y_1 + \alpha'_{22}x_2 + \alpha'_{23}x_3)x_2 \\
&\quad + (\alpha'_{31}y_1 + \alpha'_{32}x_2 + \alpha'_{33}x_3)x_3 \\
&= \alpha'_{11}y_1^2 + \alpha'_{22}x_2^2 + 2\alpha'_{23}x_2x_3 + \alpha'_{33}x_3^2;
\end{aligned}
\tag{123}
$$

here we used the symmetry of the new matrix. f has now become a quadratic form in the new independent variables y_1, x_2, x_3 and is again definite in *these* variables. [This follows from the fact that in view of (122) every triple x_1, x_2, x_3 corresponds uniquely to a triple y_1, x_2, x_3 and conversely.] But, in addition, f has a simpler form in terms of the new variables; it is now composed of a pure square $\alpha'_{11}y_1^2$ and a quadratic form

$$
f' = \alpha'_{22}x_2^2 + 2\alpha'_{23}x_2x_3 + \alpha'_{33}x_3^2
$$

in the variables x_2, x_3, i.e.,

$$
f = \alpha'_{11}y_1^2 + f'.
$$

We assert that f' is a definite form in the variables x_2, x_3. To prove this, the two conditions in 3.2 will have to be checked. For arbitrary values of x_2, x_3 and the special choice $y_1 = 0$, we find that $f' = f$. Since $f \geq 0$, it follows that $f' \geq 0$. Hence the first condition is satisfied. If $f' = 0$, then for the special choice $y_1 = 0$ we also have $f = 0$. Because of the definiteness of f as a quadratic form in y_1, x_2, x_3, this is possible only if, besides $y_1 = 0$, also $x_2 = 0$, $x_3 = 0$. The second condition is therefore also satisfied.

For f' we can repeat the treatment originally applied to f. The corresponding matrix is

$$
\begin{array}{c}
 \quad x_2 \qquad x_3 \\
\begin{array}{c} y_2' = \\ y_3' = \end{array}
\boxed{\begin{array}{cc} \alpha_{22}' & \alpha_{23}' \\ \alpha_{32}' & \alpha_{33}' \end{array}}
\end{array}
\qquad (124)
$$

where we have introduced new linear forms on the left-hand border. This matrix is obtained from (122) by omitting the first row and the first column, i.e., by omitting the new pivot row and new pivot column. In other words, it is the result of a Gauss elimination step. This matrix is again symmetric and, we have $\alpha_{22}' \neq 0$ which means that we can choose this element as pivot. The next EX-step yields

$$
\begin{array}{c}
 \quad y_2' \qquad x_3 \\
\begin{array}{c} -x_2 = \\ y_3' = \end{array}
\boxed{\begin{array}{cc} -\alpha_{22}'' & \alpha_{23}'' \\ \alpha_{32}'' & \alpha_{33}'' \end{array}}
\end{array}
$$

and, analogous to (123), we have

$$
f' = \alpha_{22}'' y_2'^2 + \alpha_{33}'' x_3^2,
$$

and therefore

$$
f = \alpha_{11}' y_1^2 + \alpha_{22}'' y_2'^2 + \alpha_{33}'' x_3^2. \qquad (124a)
$$

Hence, the quadratic form has been reduced to a sum of squares and we have obtained two important results:

1) *For the solution of symmetric-definite equations by means of the Gauss algorithm, it is possible each time to choose the element in the left-hand upper corner as pivot.*

It is never possible to hit on a zero-pivot in this way. Such equations can therefore be treated as if they had a dominant diagonal (see 1.34).

2) *A positive-definite quadratic form can be transformed into a sum of squares by means of the Gauss algorithm.*

This result can be used to determine if a given quadratic form is positive definite. In fact, if we obtain a zero-pivot during the Gauss elimination then the form is not positive definite. However, if the Gauss algorithm can be executed, then the quadratic form can be transformed into a sum of squares, (124a). Obviously the form is positive definite if and only if every one of the coefficients α'_{11}, α''_{22}, α''_{33} is strictly positive.

This result can be given a more convenient form. In fact, it follows from the exchange rules in (123) and (124a) that

$$\alpha'_{11} = \frac{1}{\alpha_{11}}, \qquad \alpha''_{22} = \frac{1}{\alpha'_{22}}.$$

Hence, α'_{11}, α''_{22}, α''_{33} are positive if and only if the three pivots α_{11}, α'_{22}, α''_{33} of the Gauss elimination are positive. This principle can be formulated generally as follows:

3) *In order to decide whether a quadratic form in n variables is positive definite, subject the corresponding symmetric matrix to the Gauss algorithm and use as pivot in each case the element in the left-hand upper corner. The form is positive definite if and only if each of the n pivots is strictly positive.*

In place of the Gauss algorithm, it is also possible to use the ordinary exchange method for the further treatment of (122). With two additional EX-steps of the modified type we can bring all x to the left border and obtain, for instance,

	y_1	y_2	y_3
$-x_1 =$	α'''_{11}	α'''_{12}	α'''_{13}
$-x_2 =$	α'''_{21}	α'''_{22}	α'''_{23}
$-x_3 =$	α'''_{31}	α'''_{32}	α'''_{33}

Again there is symmetry. Reversing the signs of all elements, we obtain the inverse matrix of the given matrix (119). It is therefore also symmetric. In addition, we find from (121) that

$$f = -\alpha'''_{11}y_1^2 - 2\alpha'''_{12}y_1y_2 - 2\alpha'''_{13}y_1y_3 - \alpha'''_{22}y_2^2 - 2\alpha'''_{23}y_2y_3 - \alpha'''_{33}y_3^2$$

and this is exactly the quadratic form corresponding to the inverse matrix. This form is therefore also positive definite.

4) *The inverse matrix of a symmetric definite matrix is again symmetric definite.*

3.31 Application to Computational Techniques

We found that for the inversion of a symmetric-definite matrix with the exchange method, as well as for the solution of a symmetric-definite system of equations with the Gauss algorithm, all matrices occurring during the computation are symmetric. Consequently, the elements below the diagonal can be written down without being calculated, and this reduces the computational work by almost one half. For the compact Gauss algorithm (see 1.34), this simplification manifests itself in the possibility of obtaining an element above the staircase lines as follows: Divide the element symmetric to it by the element directly above it in the staircase, then reverse the sign. In the notation used there, this means, for example,

$$\alpha'_{24} = -\frac{a'_{42}}{a'_{22}}; \qquad \alpha''_{34} = -\frac{a''_{43}}{a''_{33}}$$

(see also Appendix I, Example 3).

Computational methods have been developed to simplify in the symmetric case the subsequent calculation of the inverse (see 1.32). In the age of automatic computers these methods have lost some of their original importance, since they require the preparation of special computer programs, whereas the inversion can be accomplished with the simple and consistent program for the exchange method.

3.4 Orthogonality

A symmetric-definite system of equations is to be solved by means of the Gauss elimination method. According to the rectangle rule, we have for the first elimination step—that is, for the step from (119) to (124)—the following formulas:

$$\alpha'_{22} = \alpha_{22} - \frac{\alpha_{12}^2}{\alpha_{11}}, \qquad \alpha'_{33} = \alpha_{33} - \frac{\alpha_{13}^2}{\alpha_{11}}.$$

In other words: The (positive) elements on the diagonal of the original matrix are certainly decreased by the Gauss algorithm. Hence, it is possible that small (but never zero) pivot elements occur. In consideration of this phenomenon, that special case is particularly interesting where, in the given matrix (119), all elements outside the diagonal are zero and for which, therefore, the solution of the corresponding normal equations is trivial. Let us investigate further what this means when the normal equations are the result of a smoothing problem—for example, when they are the equations (103). In this case, we must have

$$(a_1, a_2) = 0, \qquad (a_1, a_3) = 0, \qquad (a_2, a_3) = 0. \tag{125}$$

Two rows of numbers are called orthogonal to each other if their scalar product is zero; (125) then states that any two of the coefficient columns of the given error equations are orthogonal to each other. *Orthogonality of the columns of the error equations therefore causes the matrix of the normal equations to be a pure diagonal matrix*—i.e., all elements in the matrix outside of the diagonal are zero. Occasionally it may be possible to arrange the measurements in such a way that the columns of the error equations are not exactly, but almost, orthogonal. More precisely, this would mean that the off-diagonal elements (a_1, a_2), (a_1, a_3), (a_2, a_3) in the normal equations (103) are small compared to the diagonal elements (a_1, a_1), (a_2, a_2), (a_3, a_3), so that the diagonal is dominant. If this dominance is strong enough, it will only be slightly lessened as a result of the Gauss algorithm; consequently, the results of the approximation will show a high degree of accuracy. In such a case we might even consider solving the normal equations iteratively (compare 1.34, last paragraph).

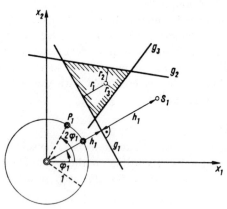

FIG. 10.   Error triangle.

Example. For the determination of a point in the x_1, x_2-plane let the m error equations

$$x_1 \cos \varphi_i + x_2 \sin \varphi_i - h_i = r_i$$

be given. The φ_i are exactly given angles while the h_i have been measured. The residuals r_i, belonging to any point of the plane, are evidently the distances of this point from the m straight lines

$$g_i: \quad x_1 \cos \varphi_i + x_2 \sin \varphi_i - h_i = 0.$$

Due to the errors in measurement, these straight lines do not intersect in one point; they form the *error figure* belonging to this approximation problem. In Fig. 10 ($m = 3$) this error figure is an *error triangle*.

The Gauss principle requires determination of the point of approximation in such a way that the sum of the squares of its distances from the straight lines g_i becomes a minimum. The angle φ_i, incidentally, is the angle between the x_1-axis and the straight line which passes through the origin and is perpendicular to g_i. The condition of orthogonality here has the form

$$\sum_{i=1}^{m} \cos \varphi_i \sin \varphi_i = 0 \qquad \text{or} \qquad \sum_{i=1}^{m} \sin 2\varphi_i = 0. \tag{126}$$

To interpret this condition geometrically, we introduce the points P_i with the coordinates $\cos 2\varphi_i$, $\sin 2\varphi_i$. They are all located on the unit circle around the origin. The condition on the right of (126) then states that the center of gravity of the points P_i must lie on the x_1-axis. Especially simple and advantageous conditions result when this center of gravity lies on the x_2-axis as well, i.e., when it is at the origin. In that case we have

$$\sum_{i=1}^{m} \cos 2\varphi_i = 0 \qquad \text{or} \qquad \sum_{i=1}^{m} \cos^2 \varphi_i = \sum_{i=1}^{m} \sin^2 \varphi_i$$

and because

$$\sum_{i=1}^{m} \cos^2 \varphi_i + \sum_{i=1}^{m} \sin^2 \varphi_i = m,$$

it follows that

$$\sum_{i=1}^{m} \cos^2 \varphi_i = \frac{m}{2}, \qquad \sum_{i=1}^{m} \sin^2 \varphi_i = \frac{m}{2}.$$

Hence, the two normal equations have the form

$$\frac{m}{2} x_1 = \sum h_i \cos \varphi_i, \qquad \frac{m}{2} x_2 = \sum h_i \sin \varphi_i. \tag{127}$$

We obtain a geometric interpretation of this situation when we introduce the points of reflection S_i of the origin relative to the straight lines g_i; they have the coordinates $2h_i \cos \varphi_i$, $2h_i \sin \varphi_i$. From (127) it follows that the approximation point is the center of gravity of these reflection points. As a concluding remark, it must be noted that the error ellipses in this special case are circles.

In general, the point of approximation in an error figure cannot be found through simple graphical construction. In the case $m = 3$, in particular, this point is usually not the center of gravity of the error triangle, nor is it any other of the geometrically significant points of the triangle. The situation is radically different in the case of the Chebyshev approximation. According to the results in 2.7, the three residuals have to be equal at the approximation point; in the case $m = 3$, this point is the center of the incircle of the error triangle.

4 Nonlinear Algebra

In the first three chapters of this book we found that for the solution of linear problems numerical mathematics provides very effective algorithms consisting of finitely many computational steps. This is altogether different for nonlinear problems. Here, mathematics can in general only improve approximate values for the solutions—obtained from another source—by adding correction terms, which in turn are solutions of an auxiliary linear problem. The generation of this auxiliary linear problem is called *linearization*; the auxiliary problem has to approximate the main problem in a certain way, still to be formulated more precisely. The *principle of correction* which we have occasionally mentioned as optional in linear algebra (see 1.33), now becomes obligatory in the nonlinear case.

Even after adding the linearized corrections, we generally fail to obtain exact solutions for the nonlinear problem. Rather, we get improved approximations which, in turn, must be used as the initial values for a new correction step. Continuing in this way, we obtain an algorithm of *successive approximations* which approximates the desired solutions increasingly more closely. The question now arises: Through how many steps must the algorithm be carried in order to obtain the result with satisfying accuracy? Nonlinear computational methods are therefore always connected with considerations of convergence and approximation error. Unfortunately, these considerations can only rarely be applied in a manner satisfactory for practical use.

At the start of a computational method of successive approximation, it is obviously sufficient to find only an approximate solution to the auxiliary linear problem for the corrections. In fact, we know that, as a result of the nonlinearity, further improvements must follow. This is an important reason for the attention paid, in linear as well as nonlinear algebra, to algorithms which deal with linear equations by means of successive approximations and which therefore yield more or less accurate solutions, depending on the number of steps used. At the end of 1.34, such an iteration method was briefly mentioned.

The great difficulty for nonlinear problems (which lies almost outside the sphere of mathematics) is the generation of the first approximating values. Initial information can be obtained from *graphical constructions* and, for many problems dealing with technical applications, from a *concrete experiment* and a *physical measurement*. In isolated cases, numerical methods exist for solving a nonlinear problem directly—that is, without knowledge of approximating values. We shall describe only two such methods in this book; namely, the solution of algebraic equations and the solution of initial-value problems for nonlinear differential equations.

4.1 Linearization

Let $f(x)$ be a function of the one independent variable x (Fig. 11). Consider

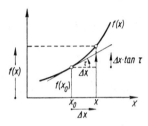

Fig. 11.   Linearization using the tangent.

the problem of approximating this function in the neighborhood of a given abscissa x_0 by a linear function. The graphical representation of this linear function will, of course, be a straight line. We select as this straight line the tangent to the given curve $f(x)$ at the point x_0. At a neighboring point x, we then have the ordinate of the tangent:

$$f(x_0) + \Delta x \tan \tau,$$

where $\Delta x = x - x_0$ is the difference between x and the given point x_0, and τ is the slope of the tangent. This ordinate value can be regarded as an approximate value of the function $f(x)$ and we therefore write

$$f(x) \sim f(x_0) + \Delta x \tan \tau.$$

The symbol $\sim$ should be read: "is approximated by." Now we know that $\tan \tau$ is the value of the derivative $f'(x)$ at the point x_0; and this leads us to the fundamental linearization formula:

$$\boxed{f(x) \sim f(x_0) + f'(x_0)\Delta x} \,. \tag{128}$$

Since x_0 is a fixed point, $f(x_0)$ and $f'(x_0)$ are also fixed numbers. Hence, the right side is a linear function of the variable Δx. Clearly, the approximation (128) will be better the closer the variable point x is to the fixed point x_0, i.e., the smaller the difference Δx. Our linearization formula assumes that the derivative $f'(x)$ exists and that it can be calculated in a simple way. If this is not the case, then we can choose another fixed abscissa X_0 in the neighborhood of x_0 and replace $f'(x_0)$ approximately by the difference quotient

$$\frac{f(X_0) - f(x_0)}{X_0 - x_0}.$$

With the abbreviation $h = X_0 - x_0$ we obtain in this way the linearization formula

$$f(x) \sim f(x_0) + [f(X_0) - f(x)] \frac{\Delta x}{h}. \tag{129}$$

The right-hand side of this formula represents a straight line which passes through the two points of the given curve $f(x)$ with the abscissas x_0, X_0 (Fig. 12). In fact, for $x = x_0$ or $\Delta x = 0$ this right side yields $f(x_0)$, and for

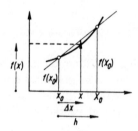

FIG. 12.   Linearization using the secant.

$x = X_0$ or $\Delta x = h$ it yields $f(X_0)$. Formula (129) is often applied to calculating a function approximately at a point x, located between the two points x_0, X_0 for which the function values $f(x_0)$, $f(X_0)$ are known (*linear interpolation*).

Following are some examples for the formula (128).

I. $f(x) = x^2$, therefore $f'(x) = 2x$.

Formula (128) yields

$$f(x) = x^2 = (x_0 + \Delta x)^2 \sim x_0^2 + 2x_0 \Delta x. \tag{130}$$

We could also have obtained this result in the following way. By expanding the square, we obtain

$$(x_0 + \Delta x)^2 = x_0^2 + 2x_0 \Delta x + (\Delta x)^2.$$

Now, we assumed that Δx is a small quantity. The square of a small quantity, however, is even smaller than the quantity itself and, as a first approximation, we can therefore neglect $(\Delta x)^2$ in relation to Δx. If we cancel $(\Delta x)^2$ in the above formula, the result is precisely (130). In other words, it is also possible to linearize by neglecting squares (and of course higher powers) of small quantities.

2. $\qquad f(x) = x^n$, therefore $f'(x) = nx^{n-1}$.

$$f(x) = x^n = (x_0 + \Delta x)^n \sim x_0^n + nx_0^{n-1}\Delta x. \qquad (131)$$

Again, we could have obtained this through binomial expansion and neglect of higher powers of (Δx).

3. The formula (131) also holds for negative and fractional exponents n. For $n = -1$, we obtain

$$\frac{1}{x_0 + \Delta x} \sim \frac{1}{x_0} - \frac{\Delta x}{x_0^2} \qquad (132)$$

(linearization of a quotient). In particular, for $x_0 = 1$ this yields the often used approximation formula

$$\frac{1}{1 + \Delta x} \sim 1 - \Delta x. \qquad (133)$$

For $|\Delta x| < 1$ the left side is the sum of the geometric series

$$1 - \Delta x + (\Delta x)^2 - (\Delta x)^3 + \dots .$$

By neglecting all powers of Δx we again obtain (133).

For $n = \frac{1}{2}$ (131) gives

$$\sqrt{x_0 + \Delta x} \sim \sqrt{x_0} + \frac{\Delta x}{2\sqrt{x_0}}. \qquad (134)$$

4. $\qquad f(x) = \sin x$, therefore $f'(x) = \cos x$.

$$f(x) = \sin x = \sin (x_0 + \Delta x) \sim \sin x_0 + \Delta x \cos x_0. \qquad (135)$$

This could have been derived as well from the addition theorem of the sine function:

$$\sin (x_0 + \Delta x) = \sin x_0 \cos \Delta x + \cos x_0 \sin \Delta x.$$

For small angles, measured in radians, the sine can be approximated by the angle and the cosine by the number 1. Hence,

$$\cos \Delta x \sim 1; \qquad \sin \Delta x \sim \Delta x.$$

Substituting this in the above formula again yields (135).

We will list several formulas here, writing simply "a" instead of x_0 and using the symbol ε instead of Δx for a small quantity.

1) $(a + \varepsilon)^2 \sim a^2 + 2a\varepsilon$ 　　　　　　2) $(a + \varepsilon)^n \sim a^n + na^{n-1}\varepsilon$

3) $\dfrac{1}{a + \varepsilon} \sim \dfrac{1}{a} - \dfrac{\varepsilon}{a^2}$ 　　　　　　4) $\sqrt{a + \varepsilon} \sim \sqrt{a} + \dfrac{\varepsilon}{2\sqrt{a}}$

5) $\sin (a + \varepsilon) \sim \sin a + \varepsilon \cos a$ 　　　　6) $\cos (a + \varepsilon) \sim \cos a - \varepsilon \sin a$

7) $e^{a+\varepsilon} \sim e^a(1 + \varepsilon)$ 　　　　　　8) $\ln (a + \varepsilon) \sim \ln a + \dfrac{\varepsilon}{a}$

$$(136)$$

(The symbol ln, of course, denotes the natural logarithm.) These formulas show us—for some common operations of mathematics—how the result is affected by a small change in the quantity to which the operation is applied.

4.11 Linearization Exercises

1. Linearize the function $f(x) = \tan x$ at the point x_0. Without using differential calculus we first obtain with the help of the formulas 5) and 6) of (136)

$$\tan (x_0 + \Delta x) = \frac{\sin (x_0 + \Delta x)}{\cos (x_0 + \Delta x)} \sim \frac{\sin x_0 + \Delta x \cos x_0}{\cos x_0 - \Delta x \sin x_0}.$$

Now we linearize the fraction

$$\frac{1}{\cos x_0 - \Delta x \sin x_0}$$

by applying formula 3) of (136) with

$$a = \cos x_0, \qquad \varepsilon = -\Delta x \sin x_0$$

(Δx is assumed to be small and hence ε is also small).

$$\frac{1}{\cos x_0 - \Delta x \sin x_0} \sim \frac{1}{\cos x_0} + \frac{\Delta x \sin x_0}{\cos^2 x_0},$$

therefore

$$\tan (x_0 + \Delta x) \sim (\sin x_0 + \Delta x \cos x_0)\left(\frac{1}{\cos x_0} + \frac{\Delta x \sin x_0}{\cos^2 x_0}\right).$$

Multiplying the parentheses and neglecting $(\Delta x)^2$ we then obtain

$$\tan (x_0 + \Delta x) \sim \tan x_0 + \Delta x(1 + \tan^2 x_0) = \tan x_0 + \frac{\Delta x}{\cos^2 x_0}.$$

2. Parallax occurring during artillery practice (Fig. 13, schematic plan), gun

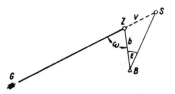

FIG. 13. Parallax during artillery practice.

in G, target at Z. The shooting is watched by an observer posted at B. The direction of the observation forms an angle ω with the direction of the shot. The distance of the observation is b. Assume that a shot hits at S, which means that its direction was perfect but that it overshot its target by the distance v. What angle ε does this length v subtend at the observer?

In the triangle the angle at S is $(\omega - \varepsilon)$. The sine theorem then yields

$$v \sin (\omega - \varepsilon) = b \sin \varepsilon,$$

which must be solved for ε. Under the assumption that ε is small, linearization then leads to

$$v (\sin \omega - \varepsilon \cos \omega) = b\varepsilon,$$

and, by solving this linear equation in ε, we obtain

$$\varepsilon = \frac{v \sin \omega}{b + v \cos \omega}.$$

v will be small compared to b (otherwise the observer could be hit easily). We can therefore simply use the formula

$$\varepsilon = \frac{v \sin \omega}{b}$$

3. Reading accuracy of a slide rule. The main scale of a slide rule normally is a logarithmic scale (Fig. 14). This means the following: When a certain subdivision of the scale is marked with the number x, the distance of this point from the left end of the scale is given by

$$y = a \log x = a\, 0.4343 \ln x.$$

Here, a is the length of the scale, $\log x$ is the logarithm of x to the base 10, and $\ln x$ the natural logarithm of x. What is the reading accuracy of the scale when the line on the runner of the slide rule can be adjusted with an accuracy of 0.2 millimeters? When x is increased by Δx, we obtain by means of formula 8) of (136) for y the change:

$$\Delta y = a \frac{0.4343}{x} \Delta x.$$

Setting $\Delta y = 0.2$ yields the desired reading accuracy

$$\Delta x = 0.4605 \frac{x}{a}.$$

It is noteworthy that the relative reading error

$$\frac{\Delta x}{x} = \frac{0.4605}{a}$$

is independent of x; in other words, it is the same along the entire slide rule. According to the above formula, for a slide rule with the length of 300 millimeters this error is 0.15%.

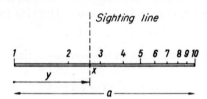

FIG. 14.   Logarithmic scale.

4.12   Functions of Several Variables

Since the functions of several variables which occur in practical problems are, for the most part, composed of functions of one variable, one usually does not need more than what we have already described.

1. Example. The function

$$f(x, y) = \sqrt{x^2 + y^2}$$

of the two variables x, y is to be linearized in the neighborhood of the given pair of values x_0, y_0. By expanding the squares and neglecting $(\Delta x)^2$ and $(\Delta y)^2$, we obtain

$$f(x, y) = \sqrt{(x_0 + \Delta x)^2 + (y_0 + \Delta y)^2} \sim \sqrt{(x_0^2 + y_0^2) + 2(x_0\Delta x + y_0\Delta y)}$$

where we have used $x = x_0 + \Delta x$ and $y = y_0 + \Delta y$. Application of formula 4) from (136) with

$$a = x_0^2 + y_0^2, \qquad \varepsilon = 2(x_0 \Delta x + y_0 \Delta y)$$

yields

$$f(x, y) = \sqrt{x^2 + y^2} \sim \sqrt{x_0^2 + y_0^2} + \frac{x_0 \Delta x + y_0 \Delta y}{\sqrt{x_0^2 + y_0^2}}. \tag{137}$$

Hence, since x_0, y_0 are fixed, a linear function of the two differences Δx, Δy has been formed here.

2. Example. $f(x, y) = xy = (x_0 + \Delta x)(y_0 + \Delta y) = x_0 y_0 + y_0 \Delta x + x_0 \Delta y + \Delta x \Delta y$. The product of two small quantities can also be neglected relative to these quantities themselves. Hence we can cancel $\Delta x \Delta y$ and so obtain the linear approximation

$$xy \sim x_0 y_0 + y_0 \Delta x + x_0 \Delta y.$$

We could also have used the formalism of differential calculus. By analogy with (128), we then find the general linearization formula for two variables

$$f(x, y) \sim f(x_0, y_0) + \frac{\partial f}{\partial x} \Delta x + \frac{\partial f}{\partial y} \Delta y \tag{138}$$

where again $x = x_0 + \Delta x$, $y = y_0 + \Delta y$.

The two partial derivatives are to be calculated for the arguments x_0, y_0.

4.2 The Correction Method of Newton

Let $f(x)$ be a given (nonlinear) function; solve the nonlinear equation $f(x) = 0$. In graphical representation (Fig. 15) this means finding those points where the curve $f(x)$ intersects the x-axis. These are called the *zeros* of the function $f(x)$, or the *roots* of the equation

$$f(x) = 0.$$

Newton's method solves this problem under the assumption that we already know one approximating value x_0 of the zero x. In this method, the function $f(x)$ is linearized in the neighborhood of x_0—i.e., it is approximated by a linear function—and then the zero x_1 of this linear function is calculated by solving a linear equation. It is thereby hoped that x_1 is a better approximation of the desired zero x than x_0. When the linearization formula (128) is used the approximating linear function has the form

$$f(x_0) + f'(x_0)\Delta x = f(x_0) + f'(x_0)(x - x_0).$$

Equating this with zero and solving for x, we find for the zero of this linear function:

$$x_1 = x_0 - \frac{f(x_0)}{f'(x_0)}. \tag{139}$$

In geometric language, this whole construction means that we are taking the tangent to the curve $f(x)$ at the point x_0 and determining its intersection x_1 with the x-axis. If we choose the linearization formula (129), we find the correction formula

$$x_1 = x_0 - \frac{f(x_0)}{f(X_0) - f(x_0)} h \tag{140}$$

(*inverse linear interpolation*). The Newton method can be iterated by again improving x_1 to become another approximating value x_2:

$$x_2 = x_1 - \frac{f(x_1)}{f'(x_1)}, \tag{141}$$

This can be continued and results in the construction of a polygon composed

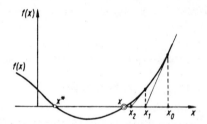

FIG. 15.   Newton method.

of vertical lines and tangents (Fig. 15); further below we will discuss the conditions for convergence of this sequence of successive approximating values $x_0, x_1, x_2, \ldots$ to the unknown zero x of $f(x)$.

1. Example. *Square-root calculation.* To calculate the square root x of a given number a, we have to solve the quadratic equation $x^2 - a = 0$; i.e., we must find the zero of the quadratic function $f(x) = x^2 - a$. Because $f'(x) = 2x$, the Newton formula (139) yields for the correction of an approximating value x_0 of the root:

$$x_1 = x_0 - \frac{x_0^2 - a}{2x_0} = \frac{1}{2}\left(x_0 + \frac{a}{x_0}\right). \tag{142}$$

This computational method is very convenient and can be executed on a desk-calculator without noting down intermediate results. One simply performs the division $q = a/x_0$ on the machine and then calculates mentally the arithmetic mean of x_0 and q. The geometric mean of these two quantities is

$$\sqrt{x_0 q} = \sqrt{x_0 \frac{a}{x_0}} = \sqrt{a},$$

i.e., the desired root x. Since the arithmetic mean of two numbers is always $\geq$ the geometric mean, we see that the approximation x_1 will certainly be too large:

$$x_1 \geq \sqrt{a}.$$

Dividing this inequality by a we obtain

$$\frac{x_1}{a} \geq \frac{1}{\sqrt{a}} \quad \text{or} \quad \frac{a}{x_1} \leq \sqrt{a}.$$

The number (a/x_1)—which, according to (142), will be needed anyway for the calculation of the next approximant x_2—is therefore a lower bound for the root. Thus, the simple algorithm (142) gives a succession of upper and lower bounds for the root, and these bounds approach each other increasingly more closely. By way of illustration, let us calculate the square root of 2.

A slide rule furnishes us with the first approximating value $x_0 = 1.41$. Then

$$\frac{a}{x_0} = \frac{2}{1.41} = 1.4185.$$

The last decimal place was rounded up, not down, in order to ensure that the arithmetic mean (142) does not become too small as a result of the rounding and will therefore not lose its property as an upper bound.

$$x_1 = \frac{1}{2}(1.41 + 1.4185) = 1.4143 \qquad \text{(rounded up)}$$

$$\frac{a}{x_1} = 1.4141 \qquad \text{(rounded down)}$$

$$1.4141 \leq \sqrt{2} \leq 1.4143.$$

A more accurate execution of the division yields

$$\frac{a}{x_1} = 1.4141272 \qquad \text{(rounded up)}.$$

Hence,

$$x_2 = \frac{1}{2}\left(x_1 + \frac{a}{x_1}\right) = 1.4142136,$$

$$\frac{a}{x_2} = 1.4142135,$$

$$\underline{\underline{1.4142135}} \leq \sqrt{2} \leq \underline{\underline{1.4142136}}.$$

2. Example. Consider the transcendental equation $\tan x = x$. Solving this equation is equivalent to the determination of the zeros of the function $f(x) = \tan x - x$. From a tangent table arranged in terms of the angle in radians [7], we obtain

$x =$	4.49	4.50
$\tan x =$	4.4223	4.6373
$f =$	−0.0677	0.1373

The unknown value x therefore lies between the two abscissas $x_0 = 4.49$ and $X_0 = 4.50$. Formula (140) yields:

$$\underline{\underline{x_1}} = 4.49 + \frac{0.0677}{0.2050}\,0.01 = \underline{\underline{4.4933}}.$$

3. Example. In theoretical astronomy and in astronautics the *Kepler equation*

$$x - e \sin x = a \tag{143}$$

plays an important role; e is the numeric eccentricity of the celestial body's path (assumed to be elliptic), a its mean anomaly and x its eccentric anomaly. To solve this equation for x we can apply the Newton rule (139) with $f(x) = x - e \sin x - a$; this results in the formula

$$x_1 = x_0 - \frac{(x_0 - e \sin x_0) - a}{1 - e \cos x_0} \tag{144}$$

for the correction of a first approximating value x_0.

This gives us a chance to discuss another method for the solution of nonlinear equations. Writing the Kepler equation in the form

$$x = a + e \sin x \tag{145}$$

the idea suggests itself of simply inserting the first approximant on the right-hand side and calculating a better approximating value on the left; i.e., of proceeding according to the rule

$$x_1 = a + e \sin x_0. \tag{146}$$

Proceeding in this way, the next approximation is given by

$$x_2 = a + e \sin x_1$$

and so on. The method will be successful if the eccentricity e is small; in other words, if the right side of the equation (145) contains the unknown x only "weakly." This method of iteration can be attempted with every equation which is given in the form

$$x = \varphi(x) \tag{147}$$

or which can be brought into this form. Here φ is a known function of x. The iteration rule then reads:

$$x_1 = \varphi(x_0), \qquad x_2 = \varphi(x_1), \qquad \text{etc.} \tag{148}$$

The convergence will be investigated below.

4.3 Recursion Formulas, Convergence

Many algorithms of numerical mathematics show the following structure: A sequence $x_0, x_1, x_2, \ldots, x_n, \ldots$ of values is calculated where in each case a value x_{n+1} is obtained from its predecessor x_n with the help of a fixed formula

$$x_{n+1} = \varphi(x_n) \tag{149}$$

(which is independent of the index n). Here φ is a given function of one variable. (149) is called a recursion formula. This algorithmic structure makes the use of *automatic digital computers* really worthwhile. In fact, the same formula must be evaluated again and again; only those numbers change to which the formula is applied. The computer program consists of the instructions for the evaluation of φ and is used once for each value of the running index n.

It can happen that the numbers x_n have an independent meaning and therefore must be included in the output of the computer program. This is the case, for example, when we are concerned with a stepwise integration of differential equations (compare 6.3). For other problems, only the limit of the sequence x_n will be of interest. The iterative solution (148) of a nonlinear equation (147) which we have just discussed is an example of such a problem.

The Newton method (139) for the solution of the equation $f(x) = 0$ is also an algorithm of the same kind. The recursion formula here reads

$$x_{n+1} = x_n - \frac{f(x_n)}{f'(x_n)},$$

i.e., we have

$$\varphi(x) = x - \frac{f(x)}{f'(x)}. \tag{150}$$

A number ω is called a *fixed point* of the recursion (149), if

$$\omega = \varphi(\omega). \tag{151}$$

The reason for this name is as follows: If we start the recursion with the special initial value $x_0 = \omega$, then (149) and (151) imply that

$$x_1 = \omega, \qquad x_2 = \omega, \ldots, \qquad x_n = \omega, \ldots,$$

in other words, the recursion remains in the fixed point. A fixed point therefore is a solution of the equation $x = \varphi(x)$, which was our original concern in (147). In the case of the Newton method we have to assume that

$$f'(\omega) \neq 0. \tag{152}$$

It then follows from (150) and (151) that $f(\omega) = 0$, i.e., that ω is a zero of $f(x)$. The hypothesis (152) requires that this zero be simple. Conversely, a simple zero of $f(x)$ is a fixed point of the Newton recursion.

We shall now investigate when and how a recursion sequence x_n converges toward a fixed point ω.

Convergence Theorem. *Let ω be a fixed point of the recursion $x_{n+1} = \varphi(x_n)$. Assume furthermore that $\varphi'(x)$ is continuous in an interval J around the point ω and that $|\varphi'(\omega)| < 1$. Then x_n converges to ω provided the initial value x_0 of the recursion is sufficiently close to ω.*

Proof. Because of $|\varphi'(\omega)| < 1$ we can find a closed subinterval J' of J where $|\varphi'(x)| < 1$ holds. Then for every point x in J' we can use the Taylor formula

$$\varphi(x) = \varphi(\omega) + \varphi'(\xi)(x - \omega)$$

where ξ is a certain point between ω and x which in general cannot be specified any further. Because of (151) this formula is equivalent to

$$\varphi(x) - \omega = \varphi'(\xi)(x - \omega). \tag{153}$$

Assume now that the starting point x_0 of the recursion is contained in J'.

Substituting this value in (153) and taking (149) into account, we obtain

$$x_1 - \omega = \varphi'(\xi_0)(x_0 - \omega), \qquad \xi_0 \text{ between } \omega \text{ and } x_0. \qquad (154)$$

ξ_0 is contained in J' and hence $|\varphi'(\xi_0)| < 1$. The equation (154) therefore states that x_1 is closer to the fixed point than x_0, and is also contained in J'. By treating x_1 and all further recursion points x_n in the same way as x_0, we find that all these points are contained in J' and that (154) can be generalized to

$$x_{n+1} - \omega = \varphi'(\xi_n)(x_n - \omega), \qquad \xi_n \text{ between } \omega \text{ and } x_n. \qquad (154a)$$

Now let $k < 1$ be the maximum of $|\varphi'(x)|$ in J'. Since ξ_n is contained in J', it follows that $|\varphi'(\xi_n)| \leq k$ and (154a) yields

$$|x_{n+1} - \omega| \leq k |x_n - \omega|,$$

and hence

$$|x_n - \omega| \leq k^n |x_0 - \omega| \to 0 \qquad \text{because } k < 1.$$

This proves the theorem. Furthermore, from $x_n \to \omega$ it now also follows that $\xi_n \to \omega$ and because of (154a) that

$$\frac{x_{n+1} - \omega}{x_n - \omega} \to \varphi'(\omega). \qquad (155)$$

In order to discuss this result we introduce the following concept:

Definition. Let $x_0, x_1, \ldots, x_n, \ldots$ be a convergent sequence (which is not necessarily the result of a recursion). We then say that the sequence *converges linearly* to the limit ω, if

$$q = \lim_{n \to \infty} \frac{x_{n+1} - \omega}{x_n - \omega} \qquad (156)$$

exists and if furthermore $|q| < 1$ but $q \neq 0$. For large n the differences $(x_n - \omega)$ are approximately equal to the terms of a geometric series with the quotient q, and therefore this value q is called the *convergence factor* of the given sequence.

Applying this to our recursion, it then follows from (155) under the conditions of the convergence theorem that the recursion converges linearly if $\varphi'(\omega) \neq 0$. The convergence factor is $\varphi'(\omega)$. Let us also discuss the case $\varphi'(\omega) = 0$. Here it is advantageous to extend the Taylor expansion by one more term:

$$\varphi(x) = \varphi(\omega) + \tfrac{1}{2}\varphi''(\xi)(x - \omega)^2.$$

Similar considerations to those above yield

$$\frac{x_{n+1} - \omega}{(x_n - \omega)^2} \to \frac{1}{2}\, \varphi''(\omega).$$

Because of the square in the denominator we speak of *quadratic convergence*. Here, (for large n), the difference $(x_{n+1} - \omega)$ is proportional to $(x_n - \omega)^2$, and not—as in the case of linear convergence—proportional only to $(x_n - \omega)$. Quadratically converging sequences therefore tend incomparably faster to their limit than linearly convergent ones.

Let us now discuss the convergence of the Newton method for the determination of a simple root of the given equation $f(x) = 0$. We have seen that ω is a fixed point of the recursion generated by the function $\varphi(x)$ in (150). It follows that

$$\varphi'(x) = \frac{f(x)f''(x)}{f'(x)^2},$$

i.e., that, at the root, $\varphi'(\omega) = 0$. The convergence theorem and our considerations about quadratic convergence yield the result:

If ω is a simple zero of the (twice continuously differentiable) function $f(x)$, then the Newton method converges quadratically to ω, provided the initial approximation x_0 is sufficiently close to ω.

Accelerating the convergence of linearly convergent sequences. Consider a sequence x_n which is linearly convergent in line with the definition (156). For large n we have approximately

$$\frac{x_{n+1} - \omega}{x_n - \omega} \sim q, \quad \text{i.e.,} \quad \omega \sim x_n + \frac{x_{n+1} - x_n}{1 - q}. \tag{157}$$

Here the right-hand side is often a much better approximating value of ω than x_n. If the convergence factor q is not already known one can try to calculate it approximately as follows: the left equation (157) yields

$$x_{n+1} - \omega \sim q(x_n - \omega)$$

and by replacing n by $(n - 1)$

$$x_n - \omega \sim q(x_{n-1} - \omega).$$

Subtraction (of these two formulas) then leads to

$$q \sim \frac{x_{n+1} - x_n}{x_n - x_{n-1}}. \tag{158}$$

An example can be found in 4.62.

The formulas (157) and (158) should not be applied casually or haphazardly since they certainly do not always bring about an acceleration of the convergence. For example, if an equation is to be solved we will first have to convince ourselves, by substituting the approximating value (157) for the solution ω, that the equation is now indeed better satisfied than it was before the convergence acceleration.

4.31 Recursions in the Case of Several Variables

We consider a computational structure of the following form

$$x_{n+1} = \varphi(x_n, y_n), \qquad y_{n+1} = \psi(x_n, y_n), \tag{159}$$

where φ, ψ are two given functions of the two variables x, y. Beginning with an initial pair of values x_0, y_0 two sequences of numbers

$$x_0, x_1, \ldots, x_n, \ldots \qquad \text{and} \qquad y_0, y_1, \ldots, y_n, \ldots$$

are generated. A pair of values (ω, η) is called a fixed point of the recursion, if

$$\omega = \varphi(\omega, \eta) \qquad \text{and} \qquad \eta = \psi(\omega, \eta). \tag{160}$$

Again we can ask whether and how the recursion sequence (159) tends to such a fixed point—i.e., whether

$$x_n \to \omega, \qquad y_n \to \eta.$$

Unfortunately, the very important discussion of this problem is not possible here with the elementary means at our disposal. Instead, we refer to the detailed monograph [9] which provides all the necessary information concerning the investigation of the convergence and the numerical stability of the recursion sequences. We will mention only that the condition $|\varphi'(\omega)| < 1$ of our convergence theorem now must be replaced by the condition that for the functional matrix

$$\begin{pmatrix} \dfrac{\partial \varphi}{\partial x} & \dfrac{\partial \varphi}{\partial y} \\[2ex] \dfrac{\partial \psi}{\partial x} & \dfrac{\partial \psi}{\partial y} \end{pmatrix},$$

(taken at the fixed point), all eigenvalues have to be < 1 in absolute value.

The iterative solution of a linear system of equations, which we mentioned briefly at the end of 1.34, belongs to this group of problems. In this special case, the recursion formulas are linear.

4.4 Newton's Method for Several Unknowns

Consider the nonlinear system of equations

$$f(x, y) = 0, \qquad g(x, y) = 0 \tag{161}$$

for the determination of two unknowns x, y; f and g are given functions of two independent variables. The Newton method proceeds analogously to the case of one unknown. We choose initial approximations x_0, y_0, introduce the corrections

$$\Delta x = x - x_0, \qquad \Delta y = y - y_0,$$

linearize f, g in the neighborhood of x_0, y_0, and solve the resulting system of linear equations for the corrections Δx, Δy. The corrected values $(x_0 + \Delta x)$ and $(y_0 + \Delta y)$ form the starting point of the next approximation, etc. We refrain from giving a theory of the method and content ourselves with presenting some examples.

I. Example. Solve the nonlinear system of equations

$$x^2 - y^2 = 1, \qquad x^2 + y^2 = 4.$$

Of course, this can be accomplished in an elementary way and without the help of approximation methods. By addition and subtraction of the two equations, we immediately find a solution

$$x = \sqrt{2.5} = \underline{\underline{1.5811}}, \qquad y = \sqrt{1.5} = \underline{\underline{1.2247}}. \tag{162}$$

By way of comparison, we also apply the Newton method. Beginning with the approximating values x_0, y_0, we set $x = x_0 + \Delta x$, $y = y_0 + \Delta y$ and linearize

$$x^2 - y^2 = (x_0 + \Delta x)^2 - (y_0 + \Delta y)^2 \sim x_0^2 - y_0^2 + 2x_0\Delta x - 2y_0\Delta y$$

$$x^2 + y^2 = (x_0 + \Delta x)^2 + (y_0 + \Delta y)^2 \sim x_0^2 + y_0^2 + 2x_0\Delta x + 2y_0\Delta y.$$

The linear equations for the corrections therefore read

$$\left. \begin{array}{l} 2x_0\Delta x - 2y_0\Delta y = 1 - (x_0^2 - y_0^2) \\[4pt] 2x_0\Delta x_0 + 2y_0\Delta y = 4 - (x_0^2 + y_0^2) \end{array} \right| \tag{163}$$

The initial approximations can best be found graphically (Fig. 16). In fact, the problem concerns the intersections of the hyperbola $x^2 - y^2 = 1$ with the circle $x^2 + y^2 = 4$. Whenever a hyperbola is involved, we can replace it in first approximation by its asymptotes. In our case, the hyperbola is equi-axial and the equation of one of the asymptotes is $y = x$.

Inserting this into the equation of the circle, we obtain the approximation point

$$x_0 = y_0 = \sqrt{2} = 1.4.$$

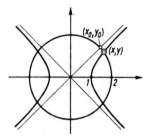

FIG. 16.   Intersections of circle and hyperbola.

The linear system of equations (163) then assumes the form

$$2.8\Delta x - 2.8\Delta y = 1$$
$$2.8\Delta x + 2.8\Delta y = 0.08$$

with the solutions $\Delta x = 0.193$, $\Delta y = -0.164$, which, in turn, leads to the improved approximation point

$$x_1 = 1.593, \qquad y_1 = 1.236.$$

The next Newton step already yields

$$x_2 = 1.5812, \qquad y_2 = 1.2248.$$

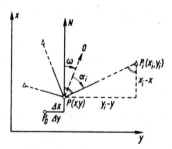

FIG. 17.   Resectioning.

2. Example. Survey of a point P with the help of the resection method (Fig. 17, situation map).   At this point P a goniometric instrument (theodolite) has been set up and bearings on certain points P_i (church steeples, trigonometric marks) have been taken.   The zero direction of the instrument forms with the direction of geographic North the (unknown) angle ω; when sighting the point P_i, we measure the angle α_i.   In the figure, the x-axis is placed in the direction of geographic North.   From the right-angled triangle shown, we obtain the relation

$$\tan (\omega + \alpha_i) = \frac{y_i - y}{x_i - x}$$

or   (164)

$$\arctan \frac{y - y_i}{x - x_i} - \omega = \alpha_i.$$

This is a nonlinear equation for the determination of the coordinates x, y of P and of the unknown orientation angle ω.   Since there are three unknowns, we have to take bearings on at least three known fixed points.   After choosing an approximation point $P_0(x_0, y_0)$ (by reading the map!) and introducing the corrections Δx, Δy, we obtain the equations

$$\arctan \frac{(y_0 - y_i) + \Delta y}{(x_0 - x_i) + \Delta x} - \omega = \alpha_i. \tag{165}$$

Linearization of the fraction by means of the formula 3) of (136) yields first

$$\frac{(y_0 - y_i) + \Delta y}{(x_0 - x_i) + \Delta x} \sim \frac{y_0 - y_i}{x_0 - x_i} + \frac{1}{(x_0 - x_i)^2} [-(y_0 - y_i)\Delta x + (x_0 - x_i)\Delta y].$$

Further, using (128), we find for the function $\arctan x$ the linearization formula

$$\arctan (x_0 + \Delta x) = \arctan x_0 + \frac{\Delta x}{1 + x_0^2}$$

and hence

$$\arctan \frac{(y_0 - y_i) + \Delta y}{(x_0 - x_i) + \Delta x} \sim \arctan \frac{y_0 - y_i}{x_0 - x_i}$$

$$+ \frac{1}{1 + \left(\dfrac{y_0 - y_i}{x_0 - x_i}\right)^2} \frac{1}{(x_0 - x_i)^2} [-(y_0 - y_i)\Delta x + (x_0 - x_i)\Delta y].$$

The linearized equations (165) therefore have the form

$$\frac{-(y_0 - y_i)\Delta x + (x_0 - x_i)\Delta y}{(x_0 - x_i)^2 + (y_0 - y_i)^2} - \omega = \alpha_i - \arctan \frac{y_0 - y_i}{x_0 - x_i}.$$

By means of the relations

$$d_i^2 = (x_i - x_0)^2 + (y_i - y_0)^2, \qquad \tan \beta_i = \frac{y_i - y_0}{x_i - x_0}$$

we introduce the azimuth angles β_i and the distances d_i from the approximation point P_0 to the fixed points P_i. Then

$$\frac{(y_i - y_0)\Delta x - (x_i - x_0)\Delta y}{d_i^2} - \omega = \alpha_i - \beta_i.$$

This linear system of equations must be solved for Δx, Δy, ω. If one has sighted more than three points, one has to smooth the measurements with the help of the method of least squares. Since the measurements of the angles have been taken with the same instrument, the absolute terms $(\alpha_i - \beta_i)$ will have the same accuracy.

4.41 Definite Nonlinear Problems

Consider the problem of finding the minimum of a given function $F(x, y)$. In 3.2 we solved this problem for the special case of a definite quadratic function $F(x, y)$. In addition, the level lines of the function were shown in Fig. 8; these are the intersections of the surface $z = F(x, y)$ with the planes parallel to the x, y-plane (z is the altitude above the x, y-plane).

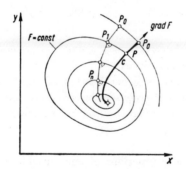

FIG. 18.   Method of steepest descent.

For illustrative purposes, we will also use this surface $z = F(x, y)$ in the case of an arbitrary function F. The given minimum problem is called *definite* if the surface is *convex*. The level lines $F(x, y) =$ const are then closed, convex curves, enclosing the unknown lowest point of the surface. In this way, we obtain Fig. 18 which can be regarded as a deformation of Fig. 8. Let us view our figure as a map of a snow-covered, bowl-shaped valley. If we tell a skier to ski down the slope of that valley to its lowest point, he is not going to use a great deal of mathematics to do it! He will, in fact, start from an abitrary point P_0, point his skis in the direction of the steepest slope and whiz downhill. Mathematically, this means the construction of that orthogonal trajectory c to all level lines which passes through the point P_0; it will finally end at the lowest point. In this way, we have sketched a very interesting method for the *direct* solution of a definite problem. We will come back to this in Chapter 6. It is called the *method of steepest descent or the gradient method*.

Right now, we will make a comment only about the use of the Newton method. When $F(x, y)$ is to become minimal, we must have for the partial derivatives

$$F_x = 0, \qquad F_y = 0.$$

Provided that the approximation point (x_0, y_0) is sufficiently close to the lowest point, we obtain the linearization

$$F_x(x_0 + \Delta x, y_0 + \Delta y) = F_x(x_0, y_0) + F_{xx}(x_0, y_0)\Delta x + F_{xy}(x_0, y_0)\Delta y$$

$$F_y(x_0 + \Delta x, y_0 + \Delta y) = F_y(x_0, y_0) + F_{yx}(x_0, y_0)\Delta x + F_{yy}(x_0, y_0)\Delta y.$$

Hence, we find for the corrections the system of linear equations

$$F_{xx}(x_0, y_0)\Delta x + F_{xy}(x_0, y_0)\Delta y + F_x(x_0, y_0) = 0$$
$$F_{yx}(x_0, y_0)\Delta x + F_{yy}(x_0, y_0)\Delta y + F_y(x_0, y_0) = 0$$

Since the mixed partial derivatives F_{xy} and F_{yx} are equal to each other, this system of equations is *symmetric*.

Finally, we note that—in generalization of our Chapter 2—we can also consider *nonlinear* (convex) *programming problems*. We will have to content ourselves here with pointing to the literature references [10], [11].

4.5 Polynomials

The simplest nonlinear functions of one variable x are the polynomials. Such a polynomial has the form

$$f(x) = P_n(x) = a_0 x^n + a_1 x^{n-1} + a_2 x^{n-2} + \ldots + a_{n-1} x + a_n; \quad (166)$$

n is an integer and the coefficients $a_0, \ldots, a_n$ are given numbers which we assume to be real. We call $P_n(x)$ a polynomial of nth degree, although it is of course possible that the *highest coefficient* a_0 is zero; i.e., that effectively only powers lower than the nth appear. A polynomial $P_1(x)$ is a linear function. If $a_0 > 0$ and $n \neq 0$, then for large positive x the polynomial will also assume large positive values. Polynomials are therefore not very well suited, for example, for the approximation of functions which are decreasing to zero. For values of x very far left on the x-axis, $P_n(x)$ will be positive or negative depending on whether n is even or odd. A polynomial of odd degree (with a highest coefficient different from zero) must therefore intersect the x-axis at least once; in other words, it has at least one real zero. A polynomial of even degree ($a_0 \neq 0$), however, does not have to possess a real root, as can be seen from the example

$$P_{2n}(x) = x^{2n} + 1.$$

The problem of determining the zeros of $P_n(x)$ leads to the equation $P_n(x) = 0$ which is called an *algebraic equation*. In the course of the centuries, a vast number of theoretical and numerical investigations have been concerned with solving such equations. But failure is often encountered in the following situation: An unknown quantity x can be represented as root of a polynomial and the coefficients of this polynomial are the results of previous calculations, i.e., they involve round-off errors. Frequently these errors influence the root x so strongly that it can be obtained only with a very unsuitable degree of approximation. Other methods for the determination of x must then be found; in 5.3 we shall have to proceed in this way when we take up eigenvalue problems. In such a situation, it is occasionally possible to alleviate the situation as follows: Instead of using the polynomial in its form (166) in terms of the powers of x, one can represent it for example in terms of *Chebyshev polynomials* (see 7.2) or by means of a set of $(n + 1)$ interpolation points (see 7.1). While the representation (166) of a polynomial is theoretically the simplest, it is rather questionable for numerical purposes.

4.51 The Horner Algorithm

In order to operate with polynomials we need some *division algorithms*. Let us first discuss the division of $P_n(x)$ by a linear polynomial of the simplest form $(x - p)$, where p is a given number. The following example illustrates how this division is usually executed in introductory courses.

Division of $P_3(x) = 2x^3 - 3x^2 + x + 5$ by $(x - 2)$

$$(2x^3 - 3x^2 + x + 5) : (x - 2) = 2x^2 + x + 3 = P_2(x)$$
$$\underline{2x^3 - 4x^2}$$
$$ \underline{x^2 + x + 5}$$
$$ \underline{x^2 - 2x}$$
$$ \underline{3x + 5}$$
$$ \underline{3x - 6}$$
$$\text{Remainder} \quad R = 11$$

A polynomial $P_2(x)$ of second degree appears as quotient and a number R as remainder; R still has to be divided by $(x - 2)$ and hence we get the equation

$$\frac{P_3(x)}{x - 2} = P_2(x) + \frac{R}{x - 2} \qquad \text{or} \qquad P_3(x) = P_2(x)(x - 2) + R.$$

Generally, the result of such a division can be written in the form of the equation

$$P_n(x) = P_{n-1}(x)(x - p) + R. \tag{167}$$

We shall now develop a simpler algorithm for the execution of this division and the evaluation of the remainder. In order to simplify the notation, we will do this for the case $n = 3$; the process runs analogously for arbitrary n. Let p and

$$P_3(x) = a_0x^3 + a_1x^2 + a_2x + a_3$$

be given. The problem is to find the quotient

$$P_2(x) = b_0x^2 + b_1x + b_2$$

and the remainder R. In this case, the equation (167) has the form

$$(a_0x^3 + a_1x^2 + a_2x + a_3) = (b_0x^2 + b_1x + b_2)(x - p) + R.$$

After multiplying the terms on the right-hand side and comparing the coefficients of equal powers of x on the right- and on the left-hand side, we find

$$a_0 = b_0, \qquad a_1 = b_1 - b_0p, \qquad a_2 = b_2 - b_1p, \qquad a_3 = R - b_2p,$$

or

$$b_0 = a_0, \qquad b_1 = a_1 + b_0p, \qquad b_2 = a_2 + b_1p, \qquad R = a_3 + b_2p.$$

These formulas give us the b and R by means of a recursive algorithm, i.e., they solve the division problem. It is convenient to perform the calculations by using the following scheme

$$
\begin{array}{c c c c c}
 & a_0 & a_1 & a_2 & a_3 \\
 & & b_0p & b_1p & b_2p \\
\hline
p) & b_0 & b_1 & b_2 & R
\end{array}
\tag{168}
$$

In the first row are the coefficients of the given $P_3(x)$. The highest coefficient is written down unchanged in the third row and yields b_0. Now we multiply by p and write the product into the second row diagonally to the right above b_0. Addition now yields b_1, and so on. In the third row, we thus obtain the coefficients of $P_2(x)$ and the remainder R. Using this Horner scheme for our numerical example above, we get

$$
\begin{array}{rrrr}
2 & -3 & 1 & 5 \\
 & 4 & 2 & 6 \\
\hline
2)\quad 2 & 1 & 3 & \underline{11}
\end{array}
$$

In particular, if we substitute $x = p$ in (167), it follows that $P_n(p) = R$. Hence, the remainder is the value of the polynomial at the point p and (167) can be written in the form

$$P_n(x) = P_{n-1}(x)(x - p) + P_n(p) \tag{169}$$

The Horner scheme offers a convenient method for calculating the value of a polynomial. In our numerical example, $P_3(2) = 11$. But also the value of the derivative at the point p can be found easily. In fact, the fundamental equation (169) can be written in the form

$$\frac{P_n(x) - P_n(p)}{x - p} = P_{n-1}(x).$$

This shows, first of all, that a *difference quotient* can also be evaluated with the help of the Horner scheme. Now let x tend to p; then

$$P'_n(p) = P_{n-1}(p). \tag{170}$$

We therefore have to evaluate $P_{n-1}(p)$, which can be done by adding another Horner scheme below the third row of the first one. Illustrating this in our numerical example we get:

$$
\begin{array}{rrrrl}
2 & -3 & 1 & 5 & \\
 & 4 & 2 & 6 & \\
\hline
2)\quad 2 & 1 & 3 & 11 & P_3(2) = 11 \\
 & 4 & 10 & & \\
\hline
2)\quad 2 & 5 & 13 & & P'_3(2) = 13
\end{array}
$$

4.52 Properties of the Roots

If p, in particular, is a root ω of the given polynomial $P_n(x)$, then the remainder $R = P_n(\omega) = 0$, the division is complete and the fundamental equation (169) reads

$$P_n(x) = P_{n-1}(x)(x - \omega). \tag{171}$$

This equation implies that every zero of $P_{n-1}(x)$ is also a zero of $P_n(x)$. Hence, we can find additional roots of $P_n(x)$ by solving the algebraic equation with lower degree, $P_{n-1}(x) = 0$. This process is called separation of the root. If, in particular, $P_{n-1}(x)$ also has the zero ω, it follows from (170) that $P'_n(\omega) = P_{n-1}(\omega) = 0$. In the graphical representation, this means that the curve $P_n(x)$ not only intersects the x-axis at the point ω, but is even tangential to it; ω is a *double* root of $P_n(x)$. All these results obviously hold true for complex roots as well.

Example for the Separation of a Root. Calculate the third roots of unity; i.e., solve the algebraic equation $x^3 = 1$, or, in other words, determine the zeros of $P_3(x) = x^3 - 1$. One root is $\omega = 1$. Separation yields:

$$
\begin{array}{cccc}
1 & 0 & 0 & -1 \\
  & 1 & 1 & 1 \\
\hline
\end{array}
$$

1) $\quad 1 \quad\quad 1 \quad\quad 1 \quad\Big| \quad 0 \quad$ = check

$$P_2(x) = x^2 + x + 1.$$

The quadratic equation $x^2 + x + 1 = 0$ has the solutions

$$\frac{-1 \pm i\sqrt{3}}{2}$$

which are also third roots of unity.

   A polynomial $P_n(x)$ can have at most n (real or complex) roots provided it is not identically equal to zero; i.e., if not all its coefficients vanish. The quickest proof for this important theorem can be obtained by means of induction. Assume that the statement has already been proved for every $P_{n-1}(x)$. (For $n = 1$ the theorem holds true.) From (171) it follows that the polynomial $P_{n-1}(x)$ obtained by division does not vanish identically. If now ω^* is a root of $P_n(x)$ different from ω, then (171) yields for $x = \omega^*$ the result

$$P_{n-1}(\omega^*)(\omega^* - \omega) = 0, \qquad \text{i.e.,} \qquad P_{n-1}(\omega^*) = 0.$$

Hence, ω^* is also a root of $P_{n-1}(x)$. According to the induction assumption, $P_{n-1}(x)$ has at most $(n - 1)$ roots and therefore $P_n(x)$ has at most $(n - 1)$ roots different from ω. This proves the theorem.

   In advanced algebra the following theorem is proved: A polynomial $P_n(x)$ which has a highest coefficient different from zero possesses *exactly* n real or complex roots provided that a root with the multiplicity m is counted as m identical roots.

4.53 Division by a Quadratic Polynomial

The theory here proceeds in the same way as for a linear divisor. But we shall be briefer. The division of $P_n(x)$ by the quadratic polynomial $(x^2 - px - q)$ yields a $P_{n-2}(x)$ as quotient and a linear polynomial $(Rx + S)$ as remainder. Hence we have the equation

$$P_n(x) = P_{n-2}(x)(x^2 - px - q) + (Rx + S). \tag{172}$$

Again denoting the coefficients of $P_n(x)$ by a_i and those of $P_{n-2}(x)$ by b_i, we easily find by comparison of coefficients the following division algorithm $(n = 6)$:

a_0	a_1	a_2	a_3	a_4	a_5	a_6
		$b_0 q$	$b_1 q$	$b_2 q$	$b_3 q$	$b_4 q$
	$b_0 p$	$b_1 p$	$b_2 p$	$b_3 p$	$b_4 p$	
b_0	b_1	b_2	b_3	b_4	R	S

Accordingly, every b-coefficient is multiplied by p and q and the resulting products are written diagonally to the right above that coefficient in the third and second rows. In the lowest row we obtain the coefficient of the polynomial $P_{n-2}(x)$ and of the remainder.

This generalized Horner scheme is needed to calculate the value of $P_n(x)$ for a given complex argument ζ. If ζ^* is the conjugate-complex value of ζ, we form the quadratic polynomial

$$(x - \zeta)(x - \zeta^*) = x^2 - px - q. \tag{173}$$

It has the *real* coefficients

$$p = \zeta + \zeta^*, \qquad q = -\zeta\zeta^* \tag{174}$$

and the roots ζ, ζ^*. Substitution of ζ in (172) therefore yields

$$P_n(\zeta) = R\zeta + S. \tag{175}$$

Only real numbers occur in the division algorithm, so that this evaluation of $P_n(\zeta)$ avoids computing with complex numbers. Because of (175), we can write (172) in the form

$$P_n(x) = P_{n-2}(x)(x^2 - px - q) + R(x - \zeta) + P_n(\zeta)$$

or

$$\frac{P_n(x) - P_n(\zeta)}{x - \zeta} = P_{n-2}(x)(x - \zeta^*) + R$$

where (173) was used. Now let x tend to ζ; then the following formula for the evaluation of the derivative results

$$P'_n(\zeta) = P_{n-2}(\zeta)(\zeta - \zeta^*) + R. \tag{176}$$

Once again, $P_{n-2}(\zeta)$ can be determined by an additional generalized Horner scheme.

In summary we can state:

Calculation of $P_n(\zeta)$, $P'_n(\zeta)$ for the complex argument $\zeta = u + iv$:

$$p = 2u, \qquad q = -(u^2 + v^2).$$

Division of $P_n(x)$ by $(x^2 - px - q)$ by means of the generalized Horner scheme yields $P_{n-2}(x)$, R, S.

$$\boxed{P_n(\zeta) = (Ru + S) + iRv; \qquad P'_n(\zeta) = R + 2ivP_{n-2}(\zeta)} \tag{177}$$

The separation of a pair of roots ζ, ζ^* can also be accomplished easily; $P_n(x)$ must then be divisible without remainder by $(x - \zeta)$ as well as by $(x - \zeta^*)$, and therefore also by (173). Hence,

$$P_n(x) = P_{n-2}(x)(x^2 - px - q)$$

and further roots can be found by solving the equation $P_{n-2}(x) = 0$.

4.54 Solution of Algebraic Equations

We show the application of the Newton method with an example of a fifth-degree equation:

$$x^5 + x + 1 = 0. \tag{178}$$

By drawing the two functions

$$y = x^5 \qquad \text{and} \qquad y = -x - 1$$

in an x, y-coordinate system and looking for their intersections, we see that the equation (178) has only one real root, namely, in the neighborhood of $x_0 = -0.8$. With or without the Horner scheme we then find for the value of the polynomial

$$P_5(x) = x^5 + x + 1$$

and of its derivative at this point

$$P_5(x_0) = -0.128, \qquad P'_5(x_0) = 3.048. \tag{179}$$

Newton correction according to formula (139) yields

$$x_1 = -0.8 + \frac{0.128}{3.048} = -0.758$$

and a further Newton correction results in

$$x_2 = -0.7549.$$

Let us substitute this value in the polynomial by means of the Horner scheme:

1	0	0	0	1	1
	-0.7549	0.56987	-0.43019	0.32475	-1.00005
1	-0.7549	0.56987	-0.43019	1.32475	-0.00005

Since, according to (179), P'_5 is about $= 3$, the underlined deviation shows that the next Newton correction amounts to less than five units of the 5th place after the decimal point. We can therefore assume that

$$\omega_1 = -0.7549$$

is the value of the first root [1] correct to four decimal places. The separation of this root has already been accomplished. On the left of the vertical line in the above scheme, we have the coefficients of the fourth degree polynomial which yields the other roots.

$$P_4(x) = x^4 - 0.7549x^3 + 0.56987x^2 - 0.43019x + 1.32475. \quad (180)$$

We know that this polynomial has only complex roots. Since we will not develop methods for finding approximate values of complex roots until later, we shall use here a special property of the polynomial $P_5(x)$, namely, that it can be factored as follows:

$$P_5(x) = x^5 + x + 1 = (x^2 + x + 1)(x^3 - x^2 + 1). \quad (181)$$

The solutions of the quadratic equation

$$x^2 + x + 1 = 0$$

therefore have to be roots of $P_4(x)$; these solutions are

$$\frac{\omega_2}{\omega_3} = \frac{-1 \pm i\sqrt{3}}{2} = -0.5 \pm 0.8660i. \quad (182)$$

[1] Methods for exact error estimation can be found in [12], Vol. 2, Chapter 6.

To practice the improvement of a complex root, we will begin the solution of the equation $P_4(x) = 0$ with the approximation value

$$\zeta_0 = -0.6 + 0.8i.$$

Using the equations (177), we then have $p = -1.2$, $q = -1$, and the two Horner schemes for the evaluation of $P_4(\zeta_0)$ and of the quotient polynomial $P_2(\zeta_0)$ read:

	1	−0.75490	0.56987	−0.43019	1.32475
−1)			−1	1.95490	−1.91575
−1.2)		−1.2	2.34588	−2.29890	
	1	−1.95490	1.91575	−0.77419	−0.59100
−1)			−1		
−1.2)		−1.2			
	1	−3.15490	0.91575		

The left-hand formula (177) gives the following result for the evaluation of $P_4(\zeta_0)$ ($R = -0.77419$, $S = -0.59100$):

$$P_4(\zeta_0) = -0.12649 - 0.61935i,$$

and for the evaluation of $P_2(\zeta_0)$ ($R = -3.15490$, $S = 0.91575$):

$$P_2(\zeta_0) = 2.80869 - 2.52392i.$$

Finally, it follows from the right-hand formula of (177) that

$$P_4'(\zeta_0) = 3.26408 + 4.49390i.$$

Using only three places after the decimal point, the Newton correction reads:

$$\zeta_1 = \zeta_0 - \frac{P_4(\zeta_0)}{P_4'(\zeta_0)} = -0.6 + 0.8i + \frac{0.126 + 0.619i}{3.264 + 4.494i} = -0.496 + 0.847i.$$

This value is already considerably closer to the correct root (182).

The separation method has a serious drawback. For example, in our case the coefficients of $P_4(x)$ are inaccurate since they depend on the value of the first root ω_1, which, in turn, has been calculated only approximately. Even a very accurately calculated root of $P_4(x)$ will therefore satisfy the equation $P_5(x) = 0$ with only a more modest degree of approximation. For that reason, the latest Newton correction should always be made with the originally given polynomial.

Now assume that after several Newton steps we have found the roots ω_2, ω_3 of (182). With $\zeta = \omega_2$, i.e.,

$$u = -0.5, \qquad v = 0.866, \qquad p = -1, \qquad q = -1,$$

this leads to the following separation:

1	−0.75490	0.56987	−0.43019	1.32475	$P_q(x)$
$q = -1)$		−1	1.75490	−1.32477	
$p = -1)$	−1	1.75490	−1.32477		
1	−1.75490	1.32477	−0.00006	−0.00002	(control!)

This, in turn, yields the quadratic equation

$$x^2 - 1.75490x + 1.32477 = 0.$$

Its solutions

$$x = 0.8774 \pm 0.7449i$$

are approximations for the two last roots ω_3, ω_4 of the equation of fifth degree. To support experiments by the reader, we here provide the seven-place values of the roots

$$\omega_1 = -0.7548777, \qquad \begin{matrix} \omega_2 \\ \omega_3 \end{matrix} = -0.5 \pm 0.8660254i,$$

$$\begin{matrix} \omega_4 \\ \omega_5 \end{matrix} = 0.8774388 \pm 0.7448617i.$$

4.6 Direct Methods for the Solution of Algebraic Equations

The improvement of rough approximate values of roots is certainly very useful. However, it is often more important for practical purposes to find, if the amount of work is not unreasonable, a three-place value for a root, than to improve it to 10-place accuracy. For that reason, we shall now develop methods yielding approximation values without knowledge from other sources.

If a polynomial $P_n(x)$ of nth degree has exactly n *real* roots[1] (which—for simplicity's sake—we shall assume to be simple), then the Newton method can be turned into a direct method.

Let x_0 be an abscissa to the right of the largest root, so that $P_n(x)$ does not have another root to the right of x_0. According to the theorem of Rolle,

---

[1] In general it is very difficult to determine whether this condition is satisfied. However, there are cases where it is known ahead of time, on the basis of the original physical problem, that all roots are real. Example: Algebraic equations belonging to symmetric (self-adjoint) eigenvalue problems (compare 5.23).

the derivative $P'_n(x)$ then has $(n-1)$ real roots, located between those of $P_n(x)$. As a polynomial of degree $(n-1)$ it also has no root to the right of x_0. By repeating this argument, we see that the same holds true for $P''_n(x)$ (and all higher derivatives). If the highest coefficient of $P_n(x)$ for example is > 0, it then follows that

$$P_n(x) > 0, \qquad P'_n(x) > 0, \qquad P''_n(x) > 0 \qquad \text{for} \qquad x > x_0.$$

Hence, to the right of x_0 the curve $P_n(x)$ is *positive, monotone increasing* and *convex*. We therefore have exactly the situation assumed in Fig. 15. When the Newton method is started with x_0, the figure shows that the next approximation x_1 lies to the left of x_0, but still to the right of the largest root x. The sequence of approximation values x_1, x_2, ... of the Newton process therefore converges monotonically to the largest root x. After separation[1] of x, a polynomial $P_{n-1}(x)$ results which again has exactly $(n-1)$ real roots. With this polynomial the procedure is repeated whereby the separated root x can be used as first approximation for the calculation of the next root x^*. In this way, the zeros are systematically rolled up from the right and none will be passed over.

To make this truly a direct method, we must still provide a rule for finding an abscissa x_0 which is to the right of the largest root. For this purpose, we can make use of the following theorem which we cite here without proof.

Theorem of Laguerre. *Let*

$$P_n(x) = x^n + a_1 x^{n-1} + \ldots + a_{n-1} x + a_n$$

be a polynomial with n real roots. Then all roots are contained in the interval whose end points are given by the two solutions of the quadratic equation

$$nx^2 + 2a_1 x + [2(n-1)a_2 - (n-2)a_1^2] = 0. \tag{183}$$

4.61 The Method of Bernoulli

This method for determining the roots of a polynomial $P_n(x)$ is founded on the following concept: Choose any auxiliary polynomial $P_m(x)$ of arbitrary degree and form the *rational function*

$$R(x) = \frac{P_m(x)}{P_n(x)}. \tag{184}$$

Then investigate the *poles* of $R(x)$, that is, those points x where $R(x)$ becomes

[1] For automatic calculation, a separation method was developed by H. Maehly [*Z. angew. Math. u. Phys.* **5** (1954) S.260] which is more advantageous than the one described.

infinite. Obviously, this can happen only if $P_n(x)$ is zero at that point.
Hence, every pole of $R(x)$ is also a zero of $P_n(x)$.[1] It is therefore possible
to replace the determination of roots of polynomials by the determination
of poles of rational functions. For this purpose, Bernoulli devised the
following method which we shall now develop under the (only inessentially
restrictive) assumption that all poles are simple. We arrange these (possibly
complex) poles in the order of increasing absolute values

$$0 < |\omega_1| \leq |\omega_2| \leq |\omega_3| \ldots \leq |\omega_\nu| \tag{185}$$

($\nu \leq n$ is the number of poles). Here we quietly assumed that the *origin
is not a pole*. Decomposition into partial fractions yields

$$R(x) = \frac{P_m(x)}{P_n(x)} = P_{m-n}(x) + \frac{A_1}{x - \omega_1} + \frac{A_2}{x - \omega_2} + \ldots + \frac{A_\nu}{x - \omega_\nu}, \tag{186}$$

The polynomial $P_{m-n}(x)$ occurs only if the degree of the numerator of $R(x)$
is $\geq$ than that of the denominator. Now $R(x)$ is expanded into a power
series. Expansion of the ith partial fraction (geometric series!) results in

$$\frac{A_i}{x - \omega_i} = -\frac{A_i}{\omega_i}\frac{1}{1 - \dfrac{x}{\omega_i}} = -A_i\left(\frac{1}{\omega_i} + \frac{x}{\omega_i^2} + \frac{x^2}{\omega_i^3} + \ldots\right).$$

We therefore introduce the constants

$$\gamma_\rho = -\frac{A_1}{\omega_1^{\rho+1}} - \frac{A_2}{\omega_2^{\rho+1}} - \ldots - \frac{A_v}{\omega_\nu^{\rho+1}}. \tag{187}$$

Then the expansion of $R(x)$ assumes the form:

$$R(x) = P_{m-n}(x) + \sum_{\rho=0}^{\infty} \gamma_\rho x^\rho. \tag{188}$$

The polynomial $P_{m-n}(x)$ interferes only in the beginning. As soon as we have
passed the power x^{m-n} in the power series, the γ_ρ are the coefficients of the
power series of $R(x)$. From (187) we obtain, after a simple transformation,

$$q_\rho = \frac{\gamma_\rho}{\gamma_{\rho-1}} = \frac{1}{\omega_1}\frac{1 + \dfrac{A_2}{A_1}\left(\dfrac{\omega_1}{\omega_2}\right)^{\rho+1} + \ldots + \dfrac{A_\nu}{A_1}\left(\dfrac{\omega_\nu}{\omega_1}\right)^{\rho+1}}{1 + \dfrac{A_2}{A_1}\left(\dfrac{\omega_1}{\omega_2}\right)^{\rho} + \ldots + \dfrac{A_\nu}{A_1}\left(\dfrac{\omega_1}{\omega_\nu}\right)^{\rho}} \tag{189}$$

[1] The reverse holds true only if $P_m(x)$ and $P_n(x)$ do not have a common zero. In
that case every root of $P_n(x)$ is also a pole of $R(x)$ and the calculation of all poles renders
all roots.

If in (185) the stronger inequality $|\omega_1| < |\omega_2|$ holds, i.e., if the absolute value of ω_1 is indeed smaller than the absolute value of all other poles, then we call ω_1 the *minimal pole*. All quotients ω_1/ω_i occuring in (189) have absolute values < 1, and by going to the limit $\rho \to \infty$ it follows that

$$\lim q_\rho = \lim \frac{\gamma_\rho}{\gamma_{\rho-1}} = \frac{1}{\omega_1}. \tag{190}$$

In order to find the minimal pole of a rational function, we have to expand it into a power series. *The quotients q_ρ of two successive coefficients of the series then tend toward the reciprocal value of the minimal pole.* Let us also investigate the speed of the convergence; in other words, let us discuss the deviation $(1/\omega_1) - q_\rho$. From (189) it follows that

$$\frac{1}{\omega_1} - q_\rho$$

$$= \frac{1}{\omega_1} \frac{\frac{A_2}{A_1}\left(\frac{\omega_1}{\omega_2}\right)^\rho\left(1 - \frac{\omega_1}{\omega_2}\right) + \frac{A_3}{A_1}\left(\frac{\omega_1}{\omega_2}\right)^\rho\left(1 - \frac{\omega_1}{\omega_3}\right) + \ldots + \frac{A_\nu}{A_1}\left(\frac{\omega_1}{\omega_\nu}\right)^\rho\left(1 - \frac{\omega_1}{\omega_\nu}\right)}{1 + \frac{A_2}{A_1}\left(\frac{\omega_1}{\omega_2}\right)^\rho + \ldots + \frac{A_\nu}{A_1}\left(\frac{\omega_1}{\omega_\nu}\right)^\rho}$$

or

$$\frac{\frac{1}{\omega_1} - q_\rho}{\left(\frac{\omega_1}{\omega_2}\right)^\rho}$$

$$= \frac{1}{\omega_1} \frac{\frac{A_2}{A_1}\left(1 - \frac{\omega_1}{\omega_2}\right) + \frac{A_3}{A_1}\left(\frac{\omega_2}{\omega_3}\right)^\rho\left(1 - \frac{\omega_1}{\omega_3}\right) + \ldots + \frac{A_\nu}{A_1}\left(\frac{\omega_2}{\omega_\nu}\right)^\rho\left(1 - \frac{\omega_1}{\omega_\nu}\right)}{1 + \frac{A_2}{A_1}\left(\frac{\omega_1}{\omega_2}\right)^\rho + \ldots + \frac{A_\nu}{A_1}\left(\frac{\omega_1}{\omega_\nu}\right)^\rho}$$

Now, if the stronger inequality $|\omega_2| < |\omega_3|$ holds in (185), then it follows, by going to the limit, that

$$\frac{\frac{1}{\omega_1} - q_\rho}{\left(\frac{\omega_1}{\omega_2}\right)^\rho} \to \frac{1}{\omega_1}\frac{A_2}{A_1}\left(1 - \frac{\omega_1}{\omega_2}\right). \tag{191}$$

By replacing ρ by $(\rho + 1)$ we obtain

$$\frac{\dfrac{1}{\omega_1} - q_{\rho+1}}{\dfrac{\omega_1}{\omega_2}\left(\dfrac{\omega_1}{\omega_2}\right)^{\rho}} \to \frac{1}{\omega_1}\frac{A_2}{A_1}\left(1 - \frac{\omega_1}{\omega_2}\right)$$

or

$$\frac{\dfrac{1}{\omega_1} - q_{\rho+1}}{\left(\dfrac{\omega_1}{\omega_2}\right)^{\rho}} \to \frac{1}{\omega_2}\frac{A_2}{A_1}\left(1 - \frac{\omega_1}{\omega_2}\right). \tag{192}$$

Division of (192) by (191) yields

$$\frac{\dfrac{1}{\omega_1} - q_{\rho+1}}{\dfrac{1}{\omega_1} - q_{\rho}} \to \frac{\omega_1}{\omega_2}.$$

The definition (156) shows the *convergence* to be *linear* with the convergence factor (ω_1/ω_2); the convergence is bad when ω_2 is close to ω_1.

Subtracting the equation (191) from (192) and introducing the differences

$$e_{\rho} = q_{\rho+1} - q_{\rho} \to 0 \tag{193}$$

we find that

$$\frac{e_{\varrho}}{\left(\dfrac{\omega_1}{\omega_2}\right)^{\rho}} \to \left(\frac{1}{\omega_1} - \frac{1}{\omega_2}\right)\frac{A_2}{A_1}\left(1 - \frac{\omega_1}{\omega_2}\right)$$

and therefore

$$\frac{e_{\rho}}{e_{\rho-1}} \to \frac{\omega_1}{\omega_2}. \tag{194}$$

Hence, the quantities e_{ρ} converge linearly to zero with the same convergence factor (ω_1/ω_2).

The equation (194) now gives us the means to calculate the second pole ω_2. For this purpose, we form the new quotients

$$q_{\rho}' = \frac{e_{\rho}}{e_{\rho-1}} q_{\rho}. \tag{195}$$

Then it follows from (190) and (194) that

$$\lim q_{\rho}' = \frac{1}{\omega_2}. \tag{196}$$

We will stop for the time being and describe later how the other poles can be found as well.

Let us summarize the rules for the computations. Consider the rational function $R(x)$ with poles ω_i which satisfy the following ordering conditions:

$$0 < |\omega_1| < |\omega_2| < |\omega_3| \leq \cdots \leq |\omega_\nu|.$$

To determine ω_1, ω_2 we expand $R(x)$ into a power series,

$$R(x) = \sum_{\rho=0}^{\infty} \gamma_\rho x^\rho$$

and form the quotients $q_\rho = \gamma_\rho/\gamma_{\rho-1}$ as well as the differences $e_\rho = q_{\rho+1} - q_\rho$ as indicated in the computational scheme of Fig. 19. The third column of the table is obtained according to rule (195):

$$q_\rho' e_{\rho-1} = e_\rho q_\rho.$$

This states that in the indicated rhombus the product of the elements on the upper right is equal to the product of the elements on the lower left. We call this rule the *first rhombus rule*. The columns q, q' in the table then tend to the reciprocal values of ω_1, ω_2, respectively.

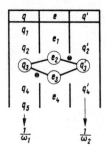

FIG. 19.   First rhombus rule.

To apply this (extended) method of Bernoulli to the solution of an *algebraic equation* $P_n(x) = 0$, we have to assume that $x = 0$ is not a solution. The constant term in $P_n(x)$ therefore is $\neq 0$. Through division by this term, we can assure that the constant term is $= 1$. For that reason, assume that $P_n(x)$ already has the form

$$P_n(x) = a_0 x^n + a_1 x^{n-1} + \cdots + a_{n-1} x + 1. \tag{197}$$

As auxiliary polynomial $P_m(x)$ in (184), it is best to take the constant 1.

For the sake of clarity we set $n = 3$. Then we have to determine the power series

$$\frac{1}{a_0 x^3 + a_1 x^2 + a_2 x + 1} = \sum \gamma_\rho x^\rho.$$

Removing the denominator we find

$$(a_0 x^3 + a_1 x^2 + a_2 x + 1)(\gamma_0 + \gamma_1 x + \gamma_2 x^2 + \gamma_3 x^3 + \gamma_4 x^4 + \dots) = 1$$

and comparison of the coefficients shows that

$$\gamma_0 = 1, \quad \gamma_1 = -a_2, \quad \gamma_2 = -a_1 - a_2 \gamma_1, \quad \gamma_3 = -a_0 - a_1 \gamma_1 - a_2 \gamma_2 \quad (198)$$

and then

$$\gamma_4 = -a_0 \gamma_1 - a_1 \gamma_2 - a_2 \gamma_3$$

and so on. All this is best evaluated by using a strip of paper on which the negative coefficients

$$-a_0, \qquad -a_1, \qquad -a_2$$

of the polynomial have been written from the top on down. The numbers γ_ρ are also calculated in a column next to which is placed the paper strip to form the scalar product (198).

Example. The cubic polynomial

$$2x^3 - 11x^2 + 13x - 4$$

has the exact roots $\omega_1 = 0.5$, $\omega_2 = 1$, $\omega_3 = 4$. In Appendix I, Example 4, the Bernoulli algorithm is worked out yielding the approximations

$$\omega_1 = 0.498, \qquad \omega_2 = 1.03.$$

This provides an opportunity to point to a phenomenon which occasionally defeats well-intended calculations in numerical mathematics. The e-column is obtained by forming the differences of the q-column. Since the q-values converge, the e-values, in time, are obtained by taking the difference of two almost equal numbers, which means that the leading decimal digits are lost and only a few significant digits remain. We call this the *loss of significant digits* through differencing. The q'-column formed from this e-column will therefore become increasingly inaccurate the further down it goes, and hence the approximation value of ω_2 suffers very heavily from round-off errors.

4.62 Accelerating Convergence

We return to the problem of determining the poles of a rational function $R(x)$. Since the q-column converges linearly, the convergence acceleration

(157) can be applied.   For the approximate convergence factor the equation (158) yields

$$q = \frac{q_{\rho+1} - q_\rho}{q_\rho - q_{\rho-1}} = \frac{e_\rho}{e_{\rho-1}} = \frac{q'_\rho}{q_\rho}.$$

Formula (157) then yields

$$\frac{1}{\omega_1} \sim q_\rho + \frac{q_{\rho+1} - q_\rho}{1 - \dfrac{q'_\rho}{q_\rho}} = q_\rho + \frac{e_\rho}{1 - \dfrac{q'_\rho}{q_\rho}} = q_\rho\left(1 + \frac{e_\rho}{q_\rho - q'_\rho}\right). \tag{199}$$

This was worked out in Appendix I, Example 4.

The following technique for accelerating the convergence is considerably more effective, but involves considerably more work as well!   As soon as we have found one approximation x_0 of the first pole ω_1, we make the following substitution in the given rational function $R(x)$ (or the given algebraic equation):

$$\xi = x - x_0. \tag{200}$$

Then we apply the Bernoulli method to the resulting rational function of ξ. Its poles are $(\omega_1 - x_0), (\omega_2 - x_0), \dots$; hence, the convergence factor in the new q-column,

$$\frac{\omega_1 - x_0}{\omega_2 - x_0},$$

is very small if x_0 was a decent approximation.   The new q-column converges rapidly.   We call this method *convergence acceleration by translation*.

4.63   Complex Poles

According to the rules of Vieta, we know that the reciprocal values of the two poles ω_1, ω_2 satisfy the quadratic equation

$$x^2 - \left(\frac{1}{\omega_1} + \frac{1}{\omega_2}\right)x + \frac{1}{\omega_1\omega_2} = 0.$$

Hence, if we form the quadratic equation

$$x^2 - (q_\rho + q'_\rho)x + q_{\rho-1}q'_\rho = 0,$$

its solutions will tend to the reciprocal values of ω_1, ω_2.   The roots of the equation

$$q_{\rho-1}q'_\rho x^2 - (q_\rho + q'_\rho)x + 1 = 0, \tag{201}$$

therefore tend to ω_1, ω_2 themselves.   We can make use of this fact when ω_1, ω_2 are close together.

Now let $R(x)$ be a rational function for which the absolutely smallest poles ω_1, ω_2 are conjugate complex; then $|\omega_1| = |\omega_2|$. In this case, the fundamental condition $|\omega_1| < |\omega_2|$ for the convergence of the Bernoulli method does not hold. The q- and q'-columns no longer converge, but show an oscillatory behavior.

Assuming that all other poles definitely have a larger absolute value than $|\omega_1|$, it is possible to show that the solutions of the quadratic equation (201) still tend to ω_1, ω_2 with increasing ρ. Appendix I, Example 5, shows the application of this to the calculation of complex roots of an algebraic equation.

4.64 The Rhombus Algorithm

Based on investigations by J. Hadamard and A. C. Aitken, H. Rutishauser developed the Bernoulli method into an algorithm which furnishes us with all the poles of a rational function, i.e., also with all the roots of a polynomial. After forming the q-, e-, q'-columns, we generate an e'-column according to the rule

$$e'_\rho = (q'_{\rho+1} - q'_\rho) + e_\rho \quad \text{or} \quad e'_\rho + q'_\rho = q'_{\rho+1} + e_\rho. \tag{202}$$

This second rhombus rule states that in the rhombus of Fig. 20 the sum

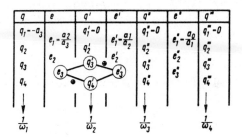

FIG. 20. Rhombus algorithm.

of the elements on the right-hand upper side must be equal to the sum of the elements on the left-hand lower side. In this way the algorithm continues: Further columns q'', e'', q''', e''', ..., are formed where for the generation of each column of the q-type the first rhombus rule is used, while for the generation of every column of the e-type the second rhombus rule applies. For the given rational function with ν poles we form altogether ν columns of the q-type (see Fig. 20). The rule $e_\rho = q_{\rho+1} - q_\rho$ for the generation of the first e-column can be considered as a special case of the second rhombus rule—we need only imagine that a column consisting entirely of zeros stands at the left beside the table of Fig. 20. The entire computational method is called the *rhombus algorithm*; it is also known as the quotient-difference algorithm or *Q-D algorithm* for short.

Now assume that the poles of the rational function satisfy the ordering relations

$$0 < |\omega_1| < |\omega_2| < \ldots < |\omega_\nu|,$$

then the *q*-columns of the rhombus scheme tend *successively* to the reciprocal poles, i.e.,

$$q_\rho \to \frac{1}{\omega_1}, \qquad q_\rho' \to \frac{1}{\omega_2}, \qquad q_\rho'' \to \frac{1}{\omega_3}, \qquad \text{etc.}$$

But if, for example, two poles occur which have the same absolute value while all other poles have either smaller or larger absolute values, then this manifests itself by the fact that two *neighboring* *q*-columns oscillate. For the purpose of determining the two troublesome poles, of equal absolute magnitude, we can form from these two columns the quadratic equation (201) and solve it. With respect to the proofs for this and other properties of the rhombus algorithm we refer to the exhaustive presentation [13].

This *column-by-column* calculation of the rhombus table is, of course, subject to the most severe loss of significant digits. To avoid this, Rutishauser also developed a rule for *row-by-row* computation. When solving an algebraic equation (197), this rule has the following form (see Fig. 20): in the first row, we enter the *q*-values

$$-a_{n-1}, \quad 0, \quad 0, \quad \ldots, \quad 0$$

(altogether *n* values). In the second row, we write the *e*-values

$$\frac{a_{n-2}}{a_{n-1}}, \qquad \frac{a_{n-3}}{a_{n-2}}, \qquad \ldots, \qquad \frac{a_1}{a_2} \qquad \frac{a_0}{a_1}. \tag{203}$$

The table (with *n* columns of the *q*-type and $(n-1)$ columns of the *e*-type) is now visualized as being extended on the left as well as on the right by a zero column. All following rows are simply calculated in such a way that the rhombus rules are satisfied (third row *q*-values, second rhombus rule; fourth row *e*-values, first rhombus rule, etc.). In this way we proceed downward until acceptable limit values for the *q*-columns appear.

If in the given algebraic equation some coefficients are missing, some of the quotients (203) might become infinite or indefinite. This can be avoided if we let an *x*-translation precede the calculations, i.e., if we perform a substitution $\xi = x - x_0$ with a suitably selected x_0. In Appendix I, Example 6, the same cubic equation as in Appendix I, Example 4, was solved through a row-by-row computation of the rhombus table. We see that the same numbers are obtained as in Example 4—only these numbers show a considerably higher degree of accuracy.

This method for the solution of algebraic equations is very convenient, especially since the preparatory power-series development is no longer necessary.

It has also been shown in [13] that the *convergence acceleration through translation* of the roots (see 4.62) can be performed by modifying the numbers of the table according to very simple rules. In addition, a method is given for the separation of already calculated roots.

4.65 Dominant Root, Stability

Consider the algebraic equation

$$P_n(x) = a_0 x^n + a_1 x^{n-1} + \ldots + a_{n-1} x + a_n = 0 \tag{204}$$

and assume that not the absolutely smallest, but the largest, the so-called dominant root of this equation has to be found. In this case we introduce the substitution $x = 1/\xi$ which yields the equation

$$a_n \xi^n + a_{n-1} \xi^{n-1} + \ldots + a_1 \xi + a_0 = 0, \tag{205}$$

where the coefficients appear in the inverse sequence; then we determine its smallest solution. If this is done with the help of the Bernoulli method, the q-column (Fig. 19) tends directly to the dominant root of (204).

When computing the circuit of an *electric oscillator* we are occasionally faced with the following problem: Let $P_n(z)$ be a polynomial in one complex variable z which ranges over the entire complex z-plane $(z = x + iy)$; it is to be determined whether $P_n(z)$ has roots with positive real part x, i.e., roots which are on the right-hand side of the imaginary axis $x = 0$. If this is the case, the corresponding oscillator or control circuit is *unstable*. However, if all roots have real parts < 0, *stability* prevails. Following a suggestion from C. Lanczos, this question can be answered as follows. The substitution

$$z = \frac{1 - \zeta}{1 + \zeta}, \quad \text{i.e.,} \quad \zeta = \frac{1 - z}{1 + z} \tag{206}$$

transforms the equation of the nth degree $P_n(z) = 0$ again into an equation of nth degree $Q_n(\zeta) = 0$ in the variable ζ; in other words, $Q_n(\zeta)$ is again a polynomial. The right-hand equation (206) maps every point of the z-plane exactly into one point of the ζ-plane. In particular, a point on the imaginary axis goes into a point on the *unit circle* of the ζ-plane (circle with center at the origin and radius $= 1$). In fact, if $z = yi$ is a point on the imaginary axis, it follows that

$$\zeta = \frac{1 - yi}{1 + yi} = \frac{(1 - y^2) - 2yi}{1 + y^2}, \qquad |\zeta|^2 = \frac{(1 - y^2)^2 + 4y^2}{(1 + y^2)^2} = 1.$$

Points to the right of the imaginary axis are mapped into points inside the unit circle (since $z = 1$ goes into $\zeta = 0$). The problem under consideration, therefore, is equivalent to the question of whether the polynomial $Q_n(\zeta)$ has roots inside the unit circle. This can be determined by using the Bernoulli method for calculating very roughly the root of $Q_n(\zeta)$ with the smallest absolute value. For other stability criteria see [1].

5 Eigenvalue Problems

The determination of frequencies in freely oscillating mechanical or electrical systems, or of critical frequencies for rotating shafts, and similar technical questions, lead to eigenvalue problems. From a mathematical viewpoint these problems belong to the field of nonlinear algebra, and we are going to develop the theory and computational technique as far as the methods of Chapter 4 will enable us. In addition, we will need some parts of Chapter 1. Eigenvalue problems for oscillatory systems of finitely many degrees of freedom can be reduced to algebraic equations, and in the following we shall describe briefly a method for this reduction as well. However, in practice this method of attack has not always proved itself; during the transition to algebraic equations, information is frequently lost due to the unavoidable computation with a limited number of digits. Therefore, methods must also be developed to solve such an eigenvalue problem directly. This requirement is a special case of the general *principle of direct attack* in numerical mathematics. While in theoretical mathematics a problem can often be simplified by means of substitutions, transformations, mappings, etc., the indiscriminate application of such methods in numerical mathematics often leads to failure. This is due to the fact that the transformed problem contains less numerical information than the original problem.

5.1 An Introductory Example

We start out from the system of springs used in 3.2 as an example (Fig. 7). There we considered two mass-points m_1, m_2 connected by springs and deflected from their equilibrium positions by the forces p_1, p_2. This time, however, we shall take three mass-points located on a horizontal line between fixed points A and B and connected with each other by four springs altogether. Then the forces p_1, p_2, p_3 are applied to these mass-points causing the displacements x_1, x_2, x_3. Using the spring constant $k = 1$ we have, instead of (116), the following equations for the displacements

$$\left.\begin{array}{rrrr} 2x_1 - & x_2 & & = p_1 \\ -x_1 + & 2x_2 - & x_3 & = p_2 \\ & - x_2 + & 2x_3 & = p_3 \end{array}\right|$$

For more than three mass-points, still further equations of the same type as the equation in the middle would have to be inserted. From this *static* problem we proceed to a *dynamic* problem by letting the system of springs oscillate freely without interference from external forces. The displacements x_1, x_2, x_3 are then functions of the time t. In the mathematical formulation of this problem, the forces p_1, p_2, p_3 must be replaced by the *forces of inertia*

$$p_1 = -m_1\ddot{x}_1, \qquad p_2 = -m_2\ddot{x}_2, \qquad p_3 = -m_3\ddot{x}_3.$$

Here, m_1, m_2, m_3 are the masses of our points and differentiation with respect to time is indicated by dots over the letters. In this way we obtain the system of differential equations

$$\begin{aligned}
2x_1 - \quad x_2 \qquad\ &= -m_1\ddot{x}_1 \\
-x_1 + 2x_2 - \quad x_3 &= -m_2\ddot{x}_2 \\
- \quad x_2 + 2x_3 &= -m_3\ddot{x}_3
\end{aligned}$$

Since that system is to perform a harmonic oscillation with the frequency ω, we now set

$$x_1 = a_1 \sin \omega t, \qquad x_2 = a_2 \sin \omega t, \qquad x_3 = a_3 \sin \omega t.$$

For the determination of the amplitudes a_1, a_2, a_3, we then obtain the linear system of equations

$$\begin{aligned}
(2 - m_1\omega^2)a_1 \qquad\quad - a_2 \qquad\qquad\quad &= 0 \\
- a_1 + (2 - m_2\omega^2)a_2 \qquad\qquad - a_3 &= 0 \\
- a_2 + (2 - m_3\omega^2)a_3 &= 0
\end{aligned}$$

Here not only a_1, a_2, a_3 are unknown, but also the frequency ω. The system of equations is *homogeneous* (see 1.2) and thus has the trivial solution $a_1 = 0$, $a_2 = 0$, $a_3 = 0$ which, however, is of no interest since physically it means a complete lack of motion in the system of springs. Our problem, rather, is to determine the frequency ω in such a way that the system of equations has a nontrivial solution. At the same time, but quite concealed, this requirement comprises the means of determining the oscillation frequencies of the system of springs. To provide a simple example for the considerations which follow, we set all point masses $= 1$, introducing the *square of the frequency* $\lambda = \omega^2$. Finally, instead of a_1, a_2, a_3 we again write x_1, x_2, x_3. This brings us to the following system of equations:

$$(2 - \lambda)x_1 \qquad - x_2 \qquad \qquad = 0$$
$$- x_1 + (2 - \lambda)x_2 \qquad - x_3 = 0$$
$$- x_2 + (2 - \lambda)x_3 = 0$$

(207)

λ must be determined in such a way that the system of equations has a *nontrivial solution*, i.e., that at least one of the unknowns x_1, x_2, x_3 is not equal to zero.

5.2   The Characteristic Polynomial

By a *special eigenvalue problem* we understand the following: We are given a linear system of equations

$$\sum_{k=1}^{n} a_{ik}x_k = \lambda x_i, \qquad i = 1, 2, \dots, n \tag{208}$$

consisting of n equations with n unknowns x_k, which on the right-hand side contain a likewise unknown parameter λ.   Bringing everything to the left-hand side, we see that the equations are homogeneous and that λ appears on the diagonal of the system.   For instance, for the case $n = 3$, and using the schematic notation of Chapter 1, this system can be written in the following form:

x_1	x_2	x_3
$a_{11} - \lambda$	a_{12}	a_{13}
a_{21}	$a_{22} - \lambda$	a_{23}
a_{31}	a_{32}	$a_{33} - \lambda$

Here the linear form in each row has to become zero.   The equations (207) form a special system of this type.

The eigenvalue problem consists in determining the parameter λ in such a way that the homogeneous equations (208) have a nontrivial solution.   We shall soon see that in general *several* values of λ exist which satisfy this requirement; they are called *eigenvalues*.   If λ is an eigenvalue then (208) has a nontrivial solution

$$x_1, x_2, \dots, x_n, \tag{209}$$

i.e., at least one of these x_k is $\neq 0$.   The row of numbers (209) is called an *eigensolution* or *eigenvector* corresponding to the eigenvalue λ.   Then too,

$$\rho x_1, \rho x_2, \dots, \rho x_n$$

for arbitrary $\rho \neq 0$ is a nontrivial solution of the homogeneous equations (208). Therefore, an eigensolution is determined *only up to a proportionality factor* ρ. It is quite possible that, corresponding to an eigenvalue λ, there exist two eigensolutions which are not proportional to each other. In that case, the eigenvalue is called *degenerate*.

The following method for the solution of an eigenvalue problem apparently goes back to R. Weber. We demonstrate it with the example of the system of three mass-points; according to (207), the eigenvalue problem reads

$$
\begin{array}{ccc}
x_1 & x_2 & x_3
\end{array}
$$

$$
\left.\begin{array}{|ccc|cc}
(2 - \lambda) & -1 & 0 & & 1 \\
-1 & (2 - \lambda) & -1 & -1 & -2 \\
0 & -1 & (2 - \lambda) & \lambda &
\end{array}\right. \qquad (210)
$$

The basic idea here is, through linear combination of these equations, to generate a new system which contains λ in the last column only. For example, if we multiply the second equation by (-1), the third equation by λ, and add, we obtain the equation

$$
x_1 - 2x_2 + (1 + 2\lambda - \lambda^2)x_3 = 0. \qquad (211)
$$

In the next step, the first two equations of (210) are multiplied by 1 and (-2), respectively, and the equation (211) by λ. Adding the three resulting equations, we find

$$
4x_1 - 5x_2 + (2 + \lambda + 2\lambda^2 - \lambda^3)x_3 = 0. \qquad (212)
$$

Together with (211) and (212), the third equation of (210) forms the new system of equations

$$
\left.\begin{array}{l}
- x_2 + (2 - \lambda)x_3 = 0 \\
x_1 - 2x_2 + (1 + 2\lambda - \lambda^2)x_3 = 0 \\
4x_1 - 5x_2 + (2 + \lambda + 2\lambda^2 - \lambda^3)x_3 = 0
\end{array}\right. \qquad (213)
$$

The requirement has thus been satisfied that λ appear in the third column only. Of course, the rules underlying these computations are still to be given; the computations here have been presented only to illustrate the basic idea. The system (213) can be written in table form as follows:

x_1	x_2		x_3		
		1	λ	λ^2	λ^3
0	-1	2	-1	0	0
$\underline{1}$	-2	1	2	-1	0
4	-5	2	1	2	-1
$x_1 =$	2	-1	-2	1	0

$$(214)$$

We now solve this system of equations by means of the Gauss algorithm where the λ-columns are treated entirely normally. The first EX-step with the underlined pivot yields

x_2		x_3		
	1	λ	λ^2	λ^3
$\underline{-1}$	2	-1	0	0
3	-2	-7	6	-1
$x_2 =$	2	-1	0	0

(The basement row was added underneath.)

The second exchange gives

x_3			
1	λ	λ^2	λ^3
4	-10	6	-1

Gathering the basement rows together with this last equation results in

$$
\begin{aligned}
x_1 &= 2x_2 + (-1 - 2\lambda + \lambda^2)x_3 \\
x_2 &= \qquad\qquad (2 - \lambda)x_3 \\
&\quad (4 - 10\lambda + 6\lambda^2 - \lambda^3)x_3 = 0
\end{aligned}
$$

$$(215)$$

The last equation can be satisfied only if

$$x_3 = 0 \qquad \text{or} \qquad 4 - 10\lambda + 6\lambda^2 - \lambda^3 = 0.$$

But substitution of $x_3 = 0$ in the first two equations yields

$$x_1 = 0, \qquad x_2 = 0.$$

We thus arrive at the trivial solution which is unsuitable. Hence, the cubic equation

$$4 - 10\lambda + 6\lambda^2 - \lambda^3 = 0 \qquad (216)$$

must be satisfied and its solutions are therefore the eigenvalues. The corresponding polynomial

$$P_3(\lambda) = \lambda^3 - 6\lambda^2 + 10\lambda - 4$$

is called the *characteristic polynomial* of the eigenvalue problem. The cubic equation has the solution $\lambda = 2$, and by separation we easily find the other two solutions. In this way, we obtain the three eigenvalues

$$\lambda_1 = 2 - \sqrt{2}, \qquad \lambda_2 = 2, \qquad \lambda_3 = 2 + \sqrt{2}$$

We still want to demonstrate the determination of the eigenvector corresponding, for example, to the eigenvalue $\lambda = 2$. Since this eigenvector is determined only up to a proportionality factor, we may assume that $x_3 = 1$. The first two equations (215) then immediately yield $x_1 = -1$, $x_2 = 0$, and hence $-1, 0, 1$ is the eigensolution. One proceeds in the same way for the other eigenvalues. Collecting the results we have:

λ	ω	x_1	x_2	x_3
0.5858	0.765	1	1.414	1
2	1.414	−1	0	1
3.4142	1.848	−1	−1.414	1

$$(217)$$

Here, ω is the frequency and the eigenvectors determine the *modes of the oscillations*.

In this example of an eigenvalue problem we considered $n = 3$ equations of the type (208) and we also found three eigenvectors. However, it is not quite true that an eigenvalue problem consisting of n equations always has n independent eigenvectors. Sometimes there are fewer (even when one admits complex eigensolutions), or sometimes more. (Such statements must always be understood in the sense that two proportional eigenvectors are not to be regarded as actually different.)

5.21 Algorithm for the Characteristic Polynomial

The computational technique for determination of the characteristic polynomial, which we have described so far only in outline, shall now be expanded to a dependable algorithm. It will be best to dovetail the elimination of λ from the first columns with the Gauss algorithm. We describe this in the following example:

$$
\begin{array}{ll}
(1) & (2 - \lambda)x_1 \quad\quad - 4x_2 \quad\quad - 2x_3 = 0 \quad\quad 2 \\
(2) & \quad - 3x_1 + (1 - \lambda)\,x_2 \quad\quad - x_3 = 0 \quad\quad 6 \quad\quad 40 \\
(3) & \quad\quad 2x_1 + \quad\quad 6x_2 + (6 - \lambda)x_3 = 0
\end{array}
$$

and the following computational table:

	x_1	x_2	x_3			
			1	λ	λ^2	λ^3
(4)	$\underline{2}$	6	6	-1			λ
(5)	-14	-2	-10	6	-1	
(6)	-120	40	-40	32	-1	-1
(7)	$x_1 =$	-3	-3	0.5	0	0
(8)		$\underline{40}$	32	-1	-1		λ
(9)		400	320	-28	-1	-1
(10)		$x_2 =$	-0.8	0.025	0.025	0
(11)			$\underline{0}$	$\underline{-18}$	$\underline{9}$	$\underline{-1}$

(218)

Here we call an equation "λ-free" if λ appears only in the last column. The equation (3) is already λ-free and is therefore carried over into the table as equation (4). We immediately prepare for a Gauss-algorithm step with this row (4) as pivot row. For this purpose we choose in this row a pivot element (in this case, the number 2); then, leaving two rows (5) and (6) free, we form the basement row (7).

In order now to construct the next λ-free equation, we multiply (4) by λ, which generates the terms 2λ and 6λ in the first two columns. There is only one possibility for eliminating these terms and that is by multiplying the equations (1) and (2) by 2 and 6 respectively and then adding the result to equation (4) after that equation has been multiplied by λ. This yields row (5) which in turn is immediately subjected to the Gauss algorithm giving the row (8). With this row, too, we prepare to continue the Gauss algorithm: Choose the pivot 40, leave row (9) free, and place the basement row into (10).

For the construction of the third λ-free equation, we multiply equation (8) just obtained, by λ, generating a term 40λ in the second column. This can be eliminated only (without causing λ-terms to develop in the first column), by adding to it equation (2) multiplied by 40. This yields row (6) which, by means of the Gauss algorithm, is transformed into rows (9) and (11). Row (11) contains the coefficients of the characteristic polynomial. The entire procedure is a sequence of unambiguous steps; table (218) contains exactly the Gauss algorithm for the solution of the linear equations (4), (5), (6) and can therefore be subjected to the usual sum controls and checks.

The mechanism of the computational procedure can be elucidated further with the help of Example 7 in Appendix I.

5.22 Exceptional Cases

The algorithm can stop early if at some step all candidates for a pivot vanish. In a special example, we shall study what happens in such a case.

(1)	$(13 - \lambda)x_1$	$- 4x_2$	$+ 2x_3 = 0$	2
(2)	$- 4\,x_1 + (13 - \lambda)x_2$		$- 2x_3 = 0$	-2
(3)	$2\,x_1$	$- 2x_2 + (10 - \lambda)x_3 = 0$		

	x_1	x_2	x_3				
				1	λ	λ^2	λ^3
(4)	2	-2		10	-1			λ
(5)	34	-34		8	10	-1	
(6)	—	—		—	—	—	
(7)	$x_1 =$	1		-5	0.5	0	
(8)		0		-162	27	-1	

The algorithm stops with row (8), since the only element in it which can serve as pivot is equal to zero.   The equation in row (8) reads:

$$(-162 + 27\lambda - \lambda^2)x_3 = 0. \tag{219}$$

By setting the first factor equal to zero we obtain a quadratic equation with the solutions $\lambda_1 = 9$, $\lambda_2 = 18$.   After substitution of $\lambda_1 = 9$ in the given equations, they are all three identical with the equation

$$2x_1 - 2x_2 + x_3 = 0. \tag{220}$$

Every triple of values x_1, x_2, x_3, which satisfies this equation, is an eigensolution corresponding to the eigenvalue $\lambda_1 = 9$.   Hence this eigenvalue is *degenerate*.   However, if we substitute $\lambda_2 = 18$ in the given equations, we obtain a homogeneous system of equations which—except for proportionality—has only the solution

$$x_1 = 2, \qquad x_2 = -2, \qquad x_3 = 1.$$

λ_2 therefore is not degenerate.   We must still consider in (219) the alternative case $x_3 = 0$.   In view of equation (3), this leads to $x_1 = x_2$ and therefore yields for example the solution 1, 1, 0.   This in turn satisfies (220) and so offers nothing new.

It can be shown that the algorithm must always stop prematurely if there are degenerate eigenvalues.   But it can also stop prematurely if none of the eigenvalues is degenerate.   In no case, however, can premature stoppage prevent a clear and exhaustive solution of the eigenvalue problem.

Only that case is dangerous where all candidates for the next pivot are not exactly zero, yet very small.   Then the further calculation can become extremely inaccurate, even when the given equations determine the eigenvalues and eigenvectors precisely.   In such a case the disturbance is caused by the method for solution of the eigenvalue problem and not by the eigenvalue problem itself.[1]   This is one of the difficulties one faces when one wants to reduce eigenvalue problems to algebraic equations.

Several different methods for generating the characteristic polynomial have been published which amount to determining *a priori* the position of the pivots in our algorithm.   This is not very advisable since the chances for occurrence of small pivots are thereby increased unduly.

5.23 Symmetric Eigenvalue Problems

A special eigenvalue problem (208) is called symmetric if the matrix of the coefficients a_{ik} is symmetric, i.e., if $a_{ik} = a_{ki}$.   (The a_{ik} are assumed to

---

[1] This is in contrast to the problem of the solution of linear systems (Chapter 1). There, all candidates for the next pivot can become small only when the given equations do not determine the unknowns precisely.

be real.)   Now let λ, λ' be two eigenvalues and x_i, x_i' the corresponding eigensolutions.   Then the equations hold

$$\sum_{k=1}^{n} a_{ik}x_k = \lambda x_i, \qquad \sum_{k=1}^{n} a_{ik}x_k' = \lambda' x_i'. \tag{221}$$

We multiply the left equation by x_i', the right by x_i, and subtract the latter from the former.   Summing over i we then have

$$\sum_{(i,k)} a_{ik}(x_i'x_k - x_k'x_i) = (\lambda - \lambda') \sum_{(i)} x_i x_i'.$$

Here the left-hand side is equal to zero since the two terms

$$a_{ik}(x_i'x_k - x_k'x_i) \qquad \text{and} \qquad a_{ki}(x_k'x_i - x_i'x_k)$$

(resulting from the interchange of i and k) cancel each other because $a_{ik} = a_{ki}$.   The important equation

$$(\lambda - \lambda') \sum_{i=1}^{n} x_i x_i' = 0 \tag{222}$$

remains.

Let us draw some conclusions from it.   First, let λ be a (real or complex) eigenvalue so that not all x_i are equal to zero.   Let λ', x_i' be the conjugate-complex values of these quantities, respectively.   Then λ' is also an eigenvalue and x_i' the corresponding eigensolution.   In fact, from the left-hand equation (221) we obtain the right one by passing to the conjugate complex values.   In this special case (222) has the form

$$(\lambda - \lambda') \sum_{i=1}^{n} |x_i|^2 = 0.$$

Here the sum is not equal to zero since not all x_i vanish.   It therefore follows that $\lambda = \lambda'$; in other words, λ is equal to its conjugate-complex value—consequently, it is real.

A symmetric eigenvalue problem can therefore have only real eigenvalues.

Secondly, let λ, λ' be two different eigenvalues; i.e., $(\lambda - \lambda') \neq 0$. It then follows from (222) that

$$\sum_{i=1}^{n} x_i x_i' = 0 \qquad \text{or} \qquad x_1 x_1' + x_2 x_2' + \ldots + x_n x_n' = 0.$$

This means that the corresponding *eigenvalues* are *orthogonal* to each other (see 3.4).   The eigenvectors of the system of springs (217) can serve as an example.   Since many technical problems lead to symmetric eigenvalue problems (compare 3.2), this result is of great importance.

5.24 The Gerschgorin Circles

It is very desirable to obtain some general information about the location of the eigenvalues of a problem (208) without having to do any extensive computations. The following theorem contributes something in this direction.

Calculate for every row of the matrix the sum of the absolute values of the coefficients a_{ik}, but leave the diagonal element out; i.e., calculate the quantities

$$r_i = |a_{i1}| + \ldots + |a_{i,i-1}| + |a_{i,i+1}| + \ldots + |a_{in}|.$$

Now draw n circles in a complex λ-plane so that the ith circle has the center a_{ii} and the radius r_i. Then all eigenvalues are contained within that domain which is covered by these disks.[1] The proof can be found in [14].

The theorem also holds when all a_{ik} are complex. On the other hand, if they are real and if there is symmetry ($a_{ik} = a_{ki}$), then, according to our considerations above, all eigenvalues are real. In this case all these eigenvalues are contained in the union of all those diameters of the Gerschgorin circles which are part of the real axis in the λ-plane. This result can be used to initiate the determination of the eigenvalues as zeros of the characteristic polynomial by means of the direct Newton method (see 4.6).

5.3 General Eigenvalue Problem, Iterative Methods

In technical applications we often encounter eigenvalue problems of the form

$$\sum_{k=1}^{n} a_{ik}x_k = \lambda \sum_{k=1}^{n} b_{ik}x_k, \qquad i = 1, 2, \ldots, n. \tag{223}$$

On the left and the right we have n linear forms in the variables $x_1, x_2, \ldots, x_n$; a_{ik} and b_{ik} are given numbers. If we bring everything to the left we obtain the homogeneous system of linear equations

$$\sum_{k=1}^{n} (a_{ik} - \lambda b_{ik})x_k = 0 \tag{224}$$

for the unknowns x_k. Every coefficient of the system of equations contains the parameter λ which must again be determined in such a way that the system of equations has a nontrivial solution. The *special* eigenvalue problem (208) appears as that special case of the *general* problem (224) where

---

[1] More precisely, this domain is the union of all these disks; it consists of all those points which are contained in the interior or on the boundary of at least one of the n circles.

λ is contained only in the coefficients on the diagonal of the system of equations. For technical applications we often have *symmetry*, i.e.,

$$a_{ik} = a_{ki} \quad \text{and} \quad b_{ik} = b_{ki}.$$

It is conceivable to try and *reduce* the general eigenvalue problem to a special one by proceeding as follows: Introduce the linear forms

$$y_i = \sum_{k=1}^{n} b_{ik} x_k$$

and invert them (see 1.2). This leads to expressions of the type

$$x_i = \sum_{k=1}^{n} \beta_{ik} y_k.$$

Substituting these linear forms in the left-hand side of (223) we obtain n linear forms in the y_k, e.g.,

$$\sum_{k=1}^{n} \alpha_{ik} y_k.$$

Problem (223) then becomes

$$\sum \alpha_{ik} y_k = \lambda y_i,$$

which is a special eigenvalue problem with the unknowns y_i. We must criticize this procedure not only because it violates the principle of direct attack, but because a possibly existent symmetry of the general problem is destroyed in this transition to the special problem.[1] In any case, these considerations show at least that a *general eigenvalue problem* can also be reduced to an *algebraic equation*.

The basic idea for the following direct attack on the general eigenvalue problem consists of considering the *inhomogeneous* equations

$$\sum_{k=1}^{n} (a_{ik} - \lambda b_{ik}) x_k = c_i \tag{225}$$

in addition to the *homogeneous* equations (224). Here the right-hand sides $c_1, c_2, \ldots, c_n$ are chosen arbitrarily. To help us understand the question a little better, we will start with a simple example. For this we shall take the

---

[1] However, a symmetry-preserving transformation to a special problem is possible by transforming the quadratic form with the coefficients b_{ik} into a sum of squares (compare 3.3).

eigenvalue problem (207) of the system of springs. The corresponding inhomogeneous equations are:

$$
\begin{vmatrix}
(2 - \lambda)x_1 & - x_2 & = c_1 \\
- x_1 + (2 - \lambda)x_2 & - x_3 = c_2 \\
- x_2 + (2 - \lambda)x_3 = c_3
\end{vmatrix}
$$

They have the solutions

$$
\begin{aligned}
x_1 &= \frac{(3 - 4\lambda + \lambda^2)c_1 + (2 - \lambda)c_2 + c_3}{4 - 10\lambda + 6\lambda^2 - \lambda^3} \\
x_3 &= \frac{c_1 + (2 - \lambda)c_2 + (3 - 4\lambda + \lambda^2)c_3}{4 - 10\lambda + 6\lambda^2 - \lambda^3}
\end{aligned}
\right\}
\tag{226}
$$

$$
x_2 = \frac{c_1 + (2 - \lambda)c_2 + c_3}{2 - 4\lambda + \lambda^2}.
\tag{227}
$$

These solutions are *rational functions* of the parameter λ and to this we can add that the numerators are linear forms in c_1, c_2, c_3. This is exactly the same in the general case (225). In fact, the solution of a system of linear equations (for example, by means of the Gauss algorithm), requires only the execution of additions, multiplications and divisions.[1] If we apply these elementary operations several times to the variable λ, a rational function of λ must necessarily result.

To avoid the crowding of indices it is convenient to use a linear combination of the unknowns

$$
\mu_1 x_1 + \mu_2 x_2 + \ldots + \mu_n x_n,
\tag{228}
$$

rather than the separate ones. Here the μ_i are any fixed chosen numbers. If we select them in such a way that their sum is $= 1$, then (228) is the mean value of the unknowns. Even without this special assumption we shall call the μ_i the *weights* and the sequence of numbers μ_i, μ_2, ... , μ_n the weight vector. Since each x_i is a rational function of λ, the combination (228) will also have to be a *rational function* of λ. We call it the *resolvent* $R(\lambda)$ of the given equations (225). By way of illustration we will form for our example the resolvent with the weights 1, 0, −1:

$$
R(\lambda) = x_1 - x_3 = (2 - 4\lambda + \lambda^2) \frac{c_1 - c_3}{4 - 10\lambda + 6\lambda^2 - \lambda^3} = \frac{c_1 - c_3}{2 - \lambda}.
\tag{229}
$$

---

[1] The mathematician would say: The solutions are contained in an algebraic field which is obtained by adjoining the indeterminate λ to the field of real numbers.

Every individual unknown x_i (considered as function of λ) is also a resolvent. In order to see this we need only take the ith weight $= 1$ and all the other weights $= 0$. Consider now any arbitrary resolvent $R(\lambda)$ of the problem (225). Then the following important theorem holds:

Theorem. *Every pole of the resolvent $R(\lambda)$ is an eigenvalue of the general eigenvalue problem* (225).

We prove this indirectly by selecting a special value λ_0 of λ which is *not* an eigenvalue and by showing that it *cannot* be a pole of the resolvent. Since λ_0 is not an eigenvalue, the homogeneous system of equations (224) has only the trivial solution for which all x_k vanish. According to the fundamental theorem of linear algebra (see 1.22), the corresponding inhomogeneous system (225) has well-determined solutions. As a result, the combination (228) also has a well-determined value which cannot be infinite. This proves the theorem. The system of springs can again serve as an illustration. We recognize, for example, in the denominator of the resolvent x_1 in (226), the characteristic polynomial of the system of springs.

Unfortunately, it is not true that conversely every eigenvalue is also a pole of the resolvent. Two counter-examples will underline this:

I. Counter-example. The resolvent (229) has only the pole $\lambda = 2$, while the system of springs has three eigenvalues which are given in (217). The fact that eigenvalues are lost here is due to an unfortunate choice of weights μ_i.

2. Counter-example. We make the special choice $c_1 = -1$, $c_2 = 0$, $c_3 = 1$ for the constants on the right side of the equations of the springs. The formulas (226) and (227) then reduce to

$$x_1 = \frac{1}{\lambda - 2}, \qquad x_2 = 0, \qquad x_3 = -\frac{1}{\lambda - 2}.$$

Every resolvent formed with arbitrary weights now has only the pole $\lambda = 2$, and two eigenvalues have been lost. The reason for this is that the chosen values of c_1, c_2, c_3 represent an eigensolution, namely the second one in the table (217).

In contrast to these results, which call for caution, we cite the following fact: Assume the constants c_i to be of the special form

$$c_i = \sum_{k=1}^{n} b_{ik}s_k. \tag{230}$$

Here the s_k are *arbitrarily chosen* numbers and we call the sequence of numbers $s_1, s_2, \ldots, s_n$ the *starting vector* for reasons which shortly will become clear.

The system of equations (225) then reads:

$$\sum_{k=1}^{n} (a_{ik} - \lambda b_{ik})x_k = \sum_{k=1}^{n} b_{ik}s_k. \tag{231}$$

Now if λ_0 is an eigenvalue of the problem (224), we can prove the following: The eigenvalue λ_0 is a pole of the resolvent, provided that the starting vector is sufficiently close to an eigenvector corresponding to λ_0, and provided that the vector of the weights $\mu_1, \mu_2, \ldots, \mu_n$ is not by chance orthogonal to this eigenvector.

To calculate the eigenvalues of the problem (224), we determine the poles of the resolvent $R(\lambda)$ by means of the *Bernoulli method* (see 4.61). In order to carry out this program, we need the power-series expansion of $R(\lambda)$. For this, in turn, we need the power series of the solutions of the system of equations (225) or (231). Assume that these latter power series have the form

$$x_k = s'_k + s''_k \lambda + s'''_k \lambda^2 + \ldots . \tag{232}$$

Substituting these series in (231) and comparing coefficients of equal powers of λ we obtain the following results:

Constant term: $\qquad \sum a_{ik}s'_k = \sum b_{ik}s_k,$ \hfill (233)

First power of λ: $\qquad \sum a_{ik}s''_k = \sum b_{ik}s'_k,$ \hfill (234)

Second power of λ: $\qquad \sum a_{ik}s'''_k = \sum b_{ik}s''_k,$ \hfill (235)

and so on. According to (233), the quantites s'_k can be found by solving a system of linear equations whose coefficients are the a_{ik} and whose right-hand sides are known. Moreover, because of (230), these right-hand sides are equal to the c_i. Equation (234) shows that the s''_k must be computed by solving a linear system of equations with *the same* coefficients a_{ik}, but with right-hand sides formed with the quantities s'_k which are now known. In this way, the process continues iteratively beginning with the starting vector s_k. We call this procedure *vector iteration*. The computational work is not as bad as it looks since, in the Gauss algorithm for the solution of the equations, only the column of the constants has to be treated anew in each case. The coefficients in the desired power series of the resolvent

$$R(\lambda) = \gamma' + \gamma''\lambda + \gamma'''\lambda^2 + \ldots \tag{236}$$

are now obtained by using (228) and applying the weights:

$$\gamma' = \sum \mu_i s'_i, \qquad \gamma'' = \sum \mu_i s''_i, \qquad \gamma''' = \sum \mu_i s'''_i, \tag{237}$$

and so on. With this we can form the first quotient column of the Bernoulli computation scheme (Fig. 19). Since the method of Bernoulli (as well as its refinement in the form of the rhombus algorithm) yields the poles of the rational function $R(\lambda)$ in order of increasing absolute values, we also obtain the eigenvalues in this same order. For oscillation problems, therefore, the frequency of the fundamental oscillation will ordinarily appear first, then that of the first harmonics, etc.[1]

We summarize the *rules for the computations* as follows:

To solve the general eigenvalue problem

$$\sum_{k=1}^{n} a_{ik}x_k = \lambda \sum_{k=1}^{n} b_{ik}x_k, \qquad i = 1, 2, \ldots n \tag{237a}$$

choose a starting vector $s_1, s_2, \ldots, s_n$ and execute the vector iteration by solving the equations (233), etc. After choosing a weight vector $\mu_1, \mu_2, \ldots, \mu_n$ form the quantities (237) and insert their quotients

$$\frac{\gamma''}{\gamma'}, \quad \frac{\gamma'''}{\gamma''}, \ldots \tag{237b}$$

into the first quotient column of the Bernoulli (or rhombus) table. Now assume that the values at the lower end of a quotient column do not change any more in their last decimal place. In that case, substitute tentatively for λ in the equations (237a) the reciprocal λ' of the lowest value and then solve these equations by means of the Gauss algorithm. This will show whether these equations are nontrivially solvable (within reasonable tolerance), i.e., whether λ' is indeed an approximate eigenvalue. At the same time, the corresponding eigensolution is obtained. Oscillating quotient columns indicate complex eigenvalues and must be treated with the quadratic equation (201).

The following supplement to the theory is also important: We assume that there exists a smallest eigenvalue λ_0; that is, that all other eigenvalues have absolute values $> |\lambda_0|$. In addition, we choose as resolvents the n solutions x_k of (231). Finally, we assume that λ_0 actually is a pole of each one of these resolvents and a simple pole in particular. We obtain n Bernoulli tables; in the kth table, the first quotient column reads

$$\frac{s_k''}{s_k'}, \quad \frac{s_k'''}{s_k''}, \ldots .$$

According to the principal result for the Bernoulli method (4.61), this column

---

[1] The rhombus scheme must be calculated columnwise and hence, as a result of the loss of significant digits mentioned in 4.61, only a few frequencies are obtained and the last ones among them very inaccurately.

has to converge toward $(1/\lambda_0)$. Merely in order to simplify the notation, we now assume that the second of the above quotients already approximates this limit very well. Then we have approximately

$$s_k'' \sim \lambda_0 s_k'''$$

and this holds for all k. Substitution in (235) yields:

$$\sum a_{ik} s_k''' \sim \lambda_0 \sum b_{ik} s_k'''.$$

This means that $s_1''', s_2''', \dots, s_n'''$ is an approximate eigensolution corresponding to λ_0. The approximation is, of course, more accurate the further the vector iteration has been carried. Hence, for oscillation problems this vector iteration also yields approximately the *mode of the fundamental oscillation*.

Finally, we shall consider the following special case. Assume that in the matrix of the coefficients a_{ik} all elements outside of the diagonal vanish and that those on the diagonal are $= 1$, i.e.,

$$a_{ik} = 0 \quad \text{for} \quad i \neq k, \quad a_{ii} = 1.$$

In this case, the given problem (223) can be written as

$$\sum_{k=1}^{n} b_{ik} x_k = \frac{1}{\lambda} x_i$$

and thus becomes a special eigenvalue problem. The iteration rule (233) then has the form:

$$s_i' = \sum_{k=1}^{n} b_{ik} s_k.$$

In other words, the iteration can be performed *without solving linear equations*. However, this attractive method for the solution of special eigenvalue problems is unsuitable for oscillation problems since it yields first the frequency of the highest harmonics (because of the transition to the reciprocal value of λ).

5.31 The Symmetric Case

Assume that $a_{ik} = a_{ki}$ and $b_{ik} = b_{ki}$. For the following, a computational rule is needed. Consider two sequences of numbers

$$u_1, u_2, \dots, u_n; \quad v_1, v_2, \dots, v_n.$$

By solving the systems of equations

$$\sum_{(k)} a_{ik} u_k' = \sum_{(k)} b_{ik} u_k; \quad \sum_{(k)} a_{ik} v_k' = \sum_{(k)} b_{ik} v_k \tag{238}$$

we subject both to one iteration step. Then the following identity between double sums holds true:

$$\sum_{(i,k)} b_{ik} u_i' v_k = \sum_{(i,k)} b_{ik} u_i v_k'. \tag{239}$$

Hence, it is possible to shift the prime from the u's to the v's (rule of the "prime shift"). *Proof:*

$$\sum_{(i,k)} b_{ik} u_i' v_k = \sum_{(i)} u_i' \sum_{(k)} b_{ik} v_k = \sum_{(i,k)} a_{ik} u_i' v_k' = \sum_{(i,k)} a_{ik} u_k' v_i'$$

$$= \sum_{(i)} v_i' \sum_{(k)} a_{ik} u_k' = \sum_{(i,k)} b_{ik} u_k v_i' = \sum_{(i,k)} b_{ik} u_i v_k'.$$

For the second equality sign, we have used the right-hand equation (238) and, for the fifth equality sign, the left-hand one. The third and sixth equality signs use the symmetry.

For symmetry problems, it is advisable to select as the weights μ_i exactly the quantities c_i of (230). The equations (237) are then modified to

$$\gamma' = \sum_{(i,k)} b_{ik} s_i' s_k, \qquad \gamma'' = \sum_{(i,k)} b_{ik} s_i'' s_k, \qquad \gamma''' = \sum_{(i,k)} b_{ik} s_i''' s_k, \tag{240}$$

and so on. It follows from the rule of the prime shift that, for example,

$$\gamma'' = \sum_{(i,k)} b_{ik} s_i' s_k'. \tag{241}$$

In other words, for the calculation of γ'' we need to iterate only once. The special choice of the weights therefore cuts down on iteration labor. By distributing the primes suitably, we obtain exactly $2m$ quantities γ with m iterations. The quotients (237b) which were formed with *these* quantities are called *Schwarz quotients*.

In the case of a *special* eigenvalue problem (208), these sums become simple sums:

$$\gamma' = \sum s_i' s_i, \qquad \gamma'' = \sum s_i'' s_i = \sum (s_i')^2, \qquad \gamma''' = \sum s_i''' s_i = \sum s_i'' s_i'. \tag{242}$$

5.32 Example for the Iteration Method

Again we take our system of springs (207), i.e., the special eigenvalue problem

$$\begin{aligned}
2x_1 - x_2 \qquad\quad &= \lambda x_1 \\
-x_1 + 2x_2 - x_3 &= \lambda x_2 \\
- x_2 + 2x_3 &= \lambda x_3
\end{aligned}$$

Obviously, the problem is *symmetric*. The iteration rule (233) here has the form

$$\begin{vmatrix} 2s_1' - s_2' & = s_1 \\ -s_1' + 2s_2' - s_3' = s_2 \\ - s_2' + 2s_3' = s_3 \end{vmatrix} \qquad (243)$$

We established earlier that the iteration method yields, with increasing number of steps, the mode of the fundamental oscillation; in view of this, we shall choose the initial deviations s_1, s_2, s_3 of our mass-points in line with our expectation of this fundamental oscillation mode.   (A transversely oscillating string gives us a better idea what to expect, rather than the longitudinally oscillating system of springs.)   Let all three mass-points be displaced in the same direction and let the middle one have the greatest displacement. Accordingly, let us try it with the starting vector

$$s_1 = 1, \qquad s_2 = 2, \qquad s_3 = 1.$$

The solution of the system of equations (243) then yields

$$s_1' = 2, \qquad s_2' = 3, \qquad s_3' = 2,$$

and after another iteration step

$$s_1'' = 3.5, \qquad s_2'' = 5, \qquad s_3'' = 3.5.$$

After division by 3.5 we obtain the displacements 1; 1.429; 1.   This already represents fairly well the fundamental oscillation which was given in (217).   Since we iterated twice, we have acquired four γ-coefficients. According to the equations (242), we find

$$\gamma' = 10, \qquad \gamma'' = 17, \qquad \gamma''' = 29, \qquad \gamma^{(4)} = \sum (s_i'')^2 = 49.5$$

and with it the following little Bernoulli table:

q	e	q'
1.7		
	0.0059	
1.7059		0.29
	0.0010	
1.7069		

By taking the reciprocal values of the lowest numbers in the q-columns, we obtain approximations for two eigenvalues:

$$\lambda_1 \sim \underline{0.5859}, \qquad \lambda_2 \sim \underline{3.4}.$$

The first one already agrees with the eigenvalue (217) of the fundamental oscillation up to one unit in the last place. Let us improve the second one by applying the method of *convergence acceleration by translation* (see 4.62). To do this, we introduce the substitution

$$\Lambda = \lambda - 3.4, \qquad \lambda = 3.4 + \Lambda \tag{244}$$

which makes the eigenvalue λ_2 small. Our eigenvalue problem then assumes the form

$$\left.\begin{aligned}
-1.4x_1 - \quad x_2 \qquad\quad &= \Lambda x_1 \\
-x_1 - 1.4x_2 - \quad x_3 &= \Lambda x_2 \\
-x_2 - 1.4x_3 &= \Lambda x_3
\end{aligned}\right|$$

The corresponding iteration rule is

$$\left.\begin{aligned}
-1.4s_1' - \quad s_2' \qquad\quad &= s_1 \\
-s_1' - 1.4s_2' - \quad s_3' &= s_2 \\
-s_2' - 1.4s_3' &= s_3
\end{aligned}\right|$$

Heuristic considerations are not very helpful in selecting the starting vector. We take an arbitrary vector approximately orthogonal to the first eigenvector, for example,

$$s_1 = 1, \qquad s_2 = -1, \qquad s_3 = 1.$$

The solution of the system of equations[1] yields

$$s_1' = 60, \qquad s_2' = -85, \qquad s_3' = 60.$$

These displacements are nearly proportional to those of the second harmonic in (217). In addition, we now have

$$\gamma' = 205, \qquad \gamma'' = 14425, \qquad \frac{\gamma''}{\gamma'} = 70.366.$$

---

[1] While solving this system, we note that small divisors occur. This disturbing development always takes place if the approximation is already close to the eigenvalue. A method of H. Unger, [1], [2], avoids this problem.

The reciprocal value 0.0142 represents an approximation for an eigenvalue Λ and according to (244) leads to the eigenvalue

$$\lambda_2 \sim 3.4142$$

of the second harmonic.

Now why does our method leave out the eigenvalue $\lambda = 2$ corresponding to the first harmonic? The answer is easy to find. All displacements occurring in our calculations have the following symmetry property. The displacement of the first mass-point is always equal to that of the third. According to (217), however, the first harmonic is antisymmetric. For it, the first displacement is opposite to the third. Hence, this first harmonic can never be approximated by our choice of starting vectors for the iteration. We formulate it mathematically as follows: Because of the special choice of the starting vectors the eigenvalue $\lambda = 2$ is not a pole of the resolvents used here. In order to find this eigenvalue one has to choose an antisymmetric starting vector.

The method of convergence acceleration by translation used here was developed by H. Wielandt and was called by him *fractional iteration*.

5.4 Outlooks

Advanced approach to eigenvalue computations proceeds in two directions. First, there are the methods adapted for use in *oscillation research*. For oscillations with a large number of degrees of freedom, these methods are aimed at calculating precisely the frequency of the fundamental oscillation and of a few harmonics. In other words, one does not want all eigenvalues, but only a few at the lower end of the spectrum. *Above all, those methods[1] should be noted here which use as tool the simultaneous iteration of several starting vectors, [1], also [15, sections 3, 4].* However, there are also variations of the above method with one starting vector, which, in many cases, satisfy the stated aim better than we have done it [16].

Second, in other fields of application the calculation of all eigenvalues may be required. Corresponding methods have been developed mainly for the special eigenvalue problem (208). In most cases, they proceed by transforming the problem in such a way that the eigenvalues are not changed but that as many coefficients a_{ik} as possible become zero in the new problem. Transformations of this kind have been given by Hessenberg, Lanczos [17] and Givens, cf. [18][2]. Rutishauer [13] has shown that, based on such transformations,

---

[1] Where no special literature references have been given in the following, consult [1], [2].

[2] Compare also A. S. Householder and F. L. Bauer: On certain methods for expanding the characteristic polynomial [*Numerische Mathematik* **1**, 29–37 (1959)], which contains many literature references.

it is possible to calculate his rhombus scheme again row by row as in the case of the solution of algebraic equations (see 4.64). This avoids the loss of significant digits. For the symmetric case, a transformation which goes back to Jacobi takes up a unique position. As a result of this transformation, all off-diagonal elements disappear in the matrix a_{ik} of the transformed problem [18]. The remaining diagonal elements are, of course, the eigenvalues.

Finally, methods also exist to separate an eigenvalue which has already been calculated. Here a new eigenvalue problem is constructed which no longer has this eigenvalue, but retains all other eigenvalues of the original problem.

Because of the large amount of computational work involved, most of these refined methods require the use of electronic computers.

5.41   Linear Oscillator

Consider a mechanical system, capable of oscillating, which is given in terms of the Lagrangian generalized coordinates $q_1, q_2, \ldots, q_n$, and is determined by its potential energy U and its kinetic energy T:

$$U = \sum_{(i,k)} a_{ik} q_i q_k, \qquad (a_{ik} = a_{ki}); \qquad T = \sum_{(i,k)} b_{ik} \dot{q}_i \dot{q}_k, \qquad (b_{ik} = b_{ki}). \quad (245)$$

Both sums are quadratic forms; the first one has as variables the generalized coordinates q_i and the second one the (generalized) velocities $\dot{q}_i$ (the dots denote differentiation with respect to time). The *Lagrange equations of motion* for this system have the form

$$\sum_{(k)} b_{ik} \ddot{q}_k + \sum_{(k)} a_{ik} q_k = 0. \quad (246)$$

Since we are interested in oscillations, we set $q_k = x_k \sin \omega t$. This immediately leads to the eigenvalue problem

$$\sum_{(k)} a_{ik} x_k = \lambda \sum_{(k)} b_{ik} x_k \quad (247)$$

where $\lambda = \omega^2$ and this is exactly our main problem (223). Hence, it is possible to give the following physical interpretation to the computational rules for the vector iteration:

First Rule.   The iteration rule (233) is obtained as follows: In the equations of motion (246) replace the *time derivatives* $\ddot{q}_k$ by the starting displacements s_k (taken negatively); furthermore, replace the Lagrange coordinates q_k by the displacements s'_k after the iteration.

Equations (240) for the coefficients of the power series of the resolvent can also be interpreted physically. For example, in order to calculate γ' we must replace, in the equation (245) for the kinetic energy, the $\dot{q}_i$ by s'_i, and the $\dot{q}_k$ by s_k. In order to dispel any doubts about this procedure, we take as an example the case of $n = 2$ degrees of freedom. Then we have for the kinetic energy:

$$T = b_{11}\dot{q}_1^2 + 2b_{12}\dot{q}_1\dot{q}_2 + b_{22}\dot{q}_2^2.$$

This must be rewritten in the following form:

$$T = b_{11}\dot{q}_1\dot{q}_1 + b_{12}\dot{q}_1\dot{q}_2$$
$$+ b_{21}\dot{q}_2\dot{q}_1 + b_{22}\dot{q}_2\dot{q}_2.$$

The mathematician calls this splitting of the middle term the *polarizing* of the quadratic form T. Only then do we replace the first factor in these products each time by an iterated displacement and the second factor by a starting displacement:

$$\gamma' = b_{11}s_1's_1 + b_{12}s_1's_2$$
$$+ b_{21}s_2's_1 + b_{22}s_2's_2.$$

In short:

Second Rule. The coefficients of the power series of the resolvent are obtained as follows: The kinetic energy is polarized and the iterated displacements of the corresponding level are inserted (the starting displacements are taken here as iterates of the 0th level).

All this can be carried over *mutatis mutandis* to oscillators with infinitely many degrees of freedom (*continuous oscillators*). We will limit ourselves to the following

Example. As shown in Fig. 21, we consider a thin cantilever beam which is fixed at its left end and is oscillating. Now the displacement is no longer

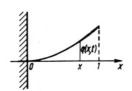

FIG. 21. Cantilever beam.

determined by a finite set of Lagrangian coordinates $q_i(t)$ but by a function $q(x, t)$ of the abscissa x and the time t. As a result, the motion is not described by a system (246) of ordinary differential equations, but by a *partial differential equation*. After suitable normalization of the material constants, this partial differential equation has the form

$$\frac{\partial^2 q}{\partial t^2} + \frac{\partial^4 q}{\partial x^4} = 0. \tag{248}$$

Together with it, we must consider the boundary conditions

$$\text{left } (x = 0): \quad q = 0, \quad q' = 0;$$
$$\text{right } (x = 1): \quad q'' = 0, \quad q''' = 0. \tag{249}$$

The primes denote derivatives with respect to x. The boundary conditions marked "left" express the fact that the beam is fixed on the left end; the boundary conditions marked "right" are the conditions for a free end as derived from the theory of elasticity. Instead of a starting vector s_i, we now have to introduce a starting function $s(x)$ which represents a first tentative displacement of the cantilever beam. Since we have used primes to denote derivatives, we can no longer signify the level of iteration by such primes; instead we will use indices. Let us therefore introduce

$$s_1(x) = \text{1st iterated function,} \qquad s_2(x) = \text{2nd iterated function,} \tag{250}$$

and so on. The first rule given above obviously leads to the iteration formula

$$-s(x) + \frac{\partial^4 s_1(x)}{\partial x^4} = 0 \quad \text{ or } \quad \frac{d^4 s_1(x)}{dx^4} = s(x). \tag{251}$$

The time variable has disappeared; with given $s(x)$, (251) is an ordinary differential equation for the function $s_1(x)$. In addition $s_1(x)$ must satisfy the boundary conditions since the iterated function is to approximate the form of the fundamental oscillation with increasing number of iterations. Altogether, we therefore obtain for the determination of $s_1(x)$ the *boundary-value problem*

$$\frac{d^4 s_1}{dx^4} = s(x), \qquad s_1(0) = s_1'(0) = 0, \qquad s_1''(1) = s_1'''(1) = 0 \tag{252}$$

and similarly we find for $s_2(x)$

$$\frac{d^4 s_2}{dx^4} = s_1(x), \qquad s_2(0) = s_2'(0) = 0, \qquad s_2''(1) = s_2'''(1) = 0. \tag{253}$$

Instead of the *system of equations* (233) in *the discontinuous case* we have in *the continuous case* a *boundary-value problem*.

The kinetic energy of the cantilever beam is

$$T = \int_0^1 \dot{q}(x, t)^2 \, dx$$

(exce_ for a factor which cancels out when we form the *Schwarz* quotients)
Hence, the second rule listed above yields for the successive γ-coefficients
(which now also carry indices instead of primes):

$$\gamma_1 = \int_0^1 s(x)s_1(x)\,dx, \qquad \gamma_2 = \int_0^1 s_1(x)^2\,dx, \qquad \gamma_3 = \int_0^1 s_1(x)s_2(x)\,dx. \qquad (254)$$

The calculation of γ_3 can be abbreviated somewhat as follows: We introduce
the *bending moment*

$$M(x) = \frac{d^2 s_2}{dx^2}. \qquad (255)$$

According to (253), it has the following properties:

$$\frac{d^2 M}{dx^2} = s_1(x), \qquad M(1) = M'(1) = 0. \qquad (256)$$

After two partial integrations we find

$$\gamma_3 = \int_0^1 M'' s_2\,dx = -\int_0^1 M' s_2'\,dx = \int_0^1 M s_2''\,dx = \int_0^1 M^2\,dx.$$

(Note that the various boundary conditions cause all integrated terms to
disappear.) With this we obtain the final equations

$$\gamma_2 = \int_0^1 s_1(x)^2\,dx, \qquad \gamma_3 = \int_0^1 M(x)^2\,dx. \qquad (257)$$

The quotient γ_2/γ_3 is an approximation for the smallest eigenvalue $\lambda = \omega^2$.
We therefore get the approximation formula

$$\omega^2 = \frac{\displaystyle\int_0^1 s_1(x)^2\,dx}{\displaystyle\int_0^1 M(x)^2\,dx} \qquad (258)$$

for the frequency of the fundamental oscillation. First approximation
formulas which differ from this one only in the consideration of the material
constants are often used in engineering for the calculation of critical numbers
of revolutions. After a starting function $s(x)$ has been chosen, $s_1(x)$ and
$M(x)$ are determined as solutions of the boundary-value problems (252) and
(256), respectively. It is of course advantageous if $s(x)$ already satisfies the

boundary conditions, but this is not necessary for convergence of the iteration procedure. In practical application, it is usual to choose the function $s_1(x)$ in such a way that the boundary conditions are satisfied and the form of the fundamental oscillation is reflected as well as possible. Then one has to solve only the boundary-value problem (256).

To illustrate this method numerically, we select as starting function simply a constant $s(x) \equiv c$. Because of the differential equation (252), $s_1(x)$ is then a polynomial of 4th degree. Its coefficients must be determined from the boundary conditions. For $c = 4! = 24$ we find

$$s_1(x) = 6x^2 - 4x^3 + x^4.$$

In the same way, we deal with the boundary value problem (256) and get

$$M(x) = (26 - 36x + 15x^4 - 6x^5 + x^6)/30.$$

Evaluation of the integrals (257) yields

$$\gamma_2 = 2.3111, \qquad \gamma_3 = 0.18691.$$

and accordingly,

$$\omega^2 = 12.36.$$

This value agrees in all four places with the exact value.

It goes without saying that our elaborations about continuous oscillators have *heuristic value* only. Next we should investigate very precisely when both our above-mentioned rules are applicable. However, we intended only to point out the connection with the exact theory of the continuous oscillators which can be found in [19], together with an abundance of applications.

6 Differential Equations

Whenever we compute numerically, we work with decimal numbers with a fixed number of places, i.e., with rational numbers. While doing this, we need to know nothing about such concepts as "real number, continuum, continuity, limits." The automatic digital computer does not know anything about these concepts either; it is organized discontinuously. All mathematical problems which use these terms—for example, which deal with functions or derivatives—have to be subjected to a process of *discretization* before they can be handled on the computer. The simplest and best known example is the replacement of a differential quotient by a difference quotient. With the discretization the problem also becomes an *algebraic* one, since now only finitely many numbers are given as input data and must be dealt with. For this, the previous parts of this book are then largely relevant. The numerical theory of differential equations and other problems of mathematical analysis will therefore consist mainly of guiding rules as to how the discretization must be carried out.

Errors will occur during this process, i.e., deviations of the discrete substitute problem from the original continuous problem. Many theoretical studies have been devoted to these errors; thanks to today's automatic computer technology, they can usually be kept small by the use of sufficiently fine discretization. Experience has shown that what happens inside the computer is more problematic. As a result of the incredibly large number of arithmetical operations executed by the computer in a very short time, all sorts of *numerical* errors occur, such as unexpected numerical instabilities, accumulations of round-off errors and loss of significant digits. In fact, here we have a real Scylla and Charybdis dilemma. The finer we make the discretization, the more operations need to be executed by the computer, and usually it is required that in the final result more decimal places be correct than in the case of a coarser discretization. It is obvious that, with the progressing technology of automatic computers, the numerical errors will increasingly outweigh the discretization errors in importance. For the field of differential equations we refer in this respect to [20].

We shall preface this chapter with a short discussion on numerical differentiation and integration. This will show that differentiation is a ticklish process and furnishes a striking example of the already mentioned Scylla and Charybdis dilemma. On the other hand, integration is a numerically stable and dependable computational process. Hence, in numerical mathematics the opposite is fortunately true from pure mathematics, where differentiation is usually simpler than integration.

6.1 Numerical Differentiation

In an x, y-coordinate system consider $(n + 1)$ abscissas $x_0, x_1, \ldots, x_n$ (called *interpolating places* or *I-places* for short) and $(n + 1)$ corresponding ordinates $y_0, y_1, \ldots, y_n$ (called *interpolating values* or *I-values*). The *I-places* shall be pairwise different from each other and numbered in order of increasing abscissas:

$$x_0 < x_1 < \ldots < x_n$$

The points $A_i(x_i, y_i)$ are named *interpolating points* (or *I-points*) and the interval from x_0 to x_n is called the *interval of the I-places*. In Fig. 22, $n = 3$

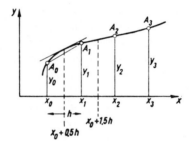

FIG. 22. Interpolation problem, differentiation.

is used and, in particular, the *I*-places are taken to be equidistant, i.e., they follow each other in constant *steps* of the length h. However, our following considerations also hold for *I*-places which are not equidistantly distributed.

The *interpolation problem* (or *I-problem*) consists in finding a polynomial of nth degree $P_n(x)$ which passes through the $(n + 1)$ *I*-points A_i. In other words, $P_n(x)$ shall be determined in such a way that

$$P_n(x_0) = y_0, \qquad P_n(x_1) = y_1, \qquad \ldots, \qquad P_n(x_n) = y_n. \tag{259}$$

We start by setting

$$P_n(x) = a_0 x^n + a_1 x^{n-1} + \ldots + a_{n-1} x + a_n$$

where the coefficients are unknown for the time being. The conditions (259) then require that

$$a_0 x_i^n + a_1 x_i^{n-1} + \ldots + a_{n-1} x_i + a_n = y_i, \qquad i = 0, 1, \ldots, n. \qquad (260)$$

This is a linear system of $(n + 1)$ equations for the determination of the $(n + 1)$ coefficients $a_0, a_1, \ldots, a_n$. Let us investigate the solvability of this system of equations by examining the corresponding homogeneous system

$$a_0 x_i^n + a_1 x_i^{n-1} + \ldots + a_{n-1} x_i + a_n = 0.$$

In this special case, all I-values are equal to zero and the equations (259) read

$$P_n(x_0) = 0, \qquad P_n(x_1) = 0, \qquad \ldots, \qquad P_n(x_n) = 0.$$

Hence, they state that the polynomial $P_n(x)$ has the $(n + 1)$ roots $x_0, x_1 \ldots, x_n$. According to a theorem in 4.52, however, a polynomial of nth degree can at most have n roots, provided it is not identically equal to zero. Since our $P_n(x)$ has $(n + 1)$ different roots, it cannot help but vanish identically. In other words, all coefficients $a_0, a_1, \ldots, a_n$ are zero. Therefore, the homogeneous system of equations possesses only the trivial solution. It then follows from the fundamental theorem of linear algebra (see 1.22) that the inhomogeneous system of equations (260) is solvable and that the solution $a_0, a_1, \ldots, a_n$ is uniquely determined. We call $P_n(x)$ the *interpolating polynomial* or *I-polynomial* for short. In summary we have the

Theorem. *For $(n + 1)$ given I-points, an I-polynomial of nth degree exists which passes through these points; it is uniquely determined.*

However, this theorem should not be interpreted to read that the power x^n actually appears in $P_n(x)$, i.e., that the highest coefficient $a_0 \neq 0$. For example, if all I-values are equal, the I-polynomial is a constant.

In practical applications, the I-values are often measured quantities. The I-polynomial can then be used to fill out the gaps between the measurements. However, it is more correct to smooth out the measuring errors by using a polynomial of lower degree and employing a least square approximation as in the fourth example in 3.12.

Assume now that the I-values are the function values of a given continuous function $f(x)$:

$$y_i = f(x_i). \qquad (261)$$

Assume further that all derivatives of $f(x)$ which occur in the following considerations exist and are continuous. We shall try to calculate the values of these derivatives approximately from the given I-values (261), (Fig. 22). To do this, we make the assumption that the I-places are equidistant with the step length h, i.e.

$$x_1 = x_0 + h, \qquad x_2 = x_0 + 2h, \qquad x_3 = x_0 + 3h, \ldots.$$

This is an assumption which is usually satisfied for tabulated functions.

As value of the derivative $f'(x)$ in the center point $(x_0 + 0.5h)$ of the first interval, we can take approximately the difference quotient

$$f'(x_0 + 0.5h) \sim \frac{y_1 - y_0}{h} \quad \text{and analogously} \quad f'(x_0 + 1.5h) \sim \frac{y_2 - y_1}{h}. \quad (262)$$

Incidentally, these formulas are exact if $f(x)$ is a quadratic polynomial, i.e., is represented by a parabola. This follows from the fact that the tangent to the parabola at $(x_0 + 0.5h)$ is parallel to the secant A_0A_1. The equations (262) yield two I-values of the function $f'(x)$ corresponding to the I-places $(x_0 + 0.5h)$, $(x_0 + 1.5h)$ which have x_1 as center point. By treating $f'(x)$ in the same way as we did $f(x)$, we find

$$f''(x_1) \sim \frac{f'(x_0 + 1.5h) - f'(x_0 + 0.5h)}{h} \sim \frac{y_2 - 2y_1 + y_0}{h^2} \quad (263)$$

where the values (262) have been used. Similarly,

$$f''(x_2) \sim \frac{y_3 - 2y_2 + y_1}{h^2} \quad (264)$$

holds.

Repetition of this process then yields

$$f'''(x_0 + 1.5h) \sim \frac{f''(x_2) - f''(x_1)}{h} \sim \frac{y_3 - 3y_2 + 3y_1 - y_0}{h^3}. \quad (265)$$

The rule for generating the absolute values of the coefficients in these formulas obviously can be written in the following condensed form:

$$
\begin{array}{ccccc}
1 & 1 & & & \\
 & 1 & 1 & & \\
\hline
1 & 2 & 1 & & \\
 & 1 & 2 & 1 & \\
\hline
1 & 3 & 3 & 1 & \\
 & 1 & 3 & 3 & 1 \\
\hline
1 & 4 & 6 & 4 & 1
\end{array}
$$

This is a slightly modified representation of the well-known Pascal triangle. Hence, the coefficients are the *binomial coefficients* with alternating signs, the coefficient of the I-value with the highest index always being equal to $+1$. In summary:

Rule for numerical differentiation. Assume we know $(m + 1)$ equidistant I-values of a function. Then the value of its mth derivative at the center of the interval of the corresponding $(m + 1)$ I-places can be calculated approximately as follows: With the alternating binomial coefficients a linear combination of the I-values is formed which is then divided by h^m ($h = $ step length).

(For example, in the formulas (263), (264), $m = 2$ was used.)

We now come to the discussion of the *discretization error*. As a representative example of such a discussion we shall concern ourselves with formula (265). Let

$$P_3(x) = a_0 x^3 + a_1 x^2 + a_2 x + a_3$$

be an *arbitrary* polynomial of third degree. In order to simplify the following calculation, we may assume, without restricting the generality, that $x_0 = 0$, i.e.,

$$x_1 = h, \qquad x_2 = 2h, \qquad x_3 = 3h.$$

The values of the polynomial at the I-places then are

$$
\begin{array}{llr}
P_3(x_0) = & a_3 & -1 \\
P_3(x_1) = & a_0 h^3 + a_1 h^2 + a_2 h + a_3 & 3 \\
P_3(x_2) = & 8a_0 h^3 + 4a_1 h^2 + 2a_2 h + a_3 & -3 \\
P_3(x_3) = & 27a_0 h^3 + 9a_1 h^2 + 3a_2 h + a_3 & 1
\end{array}
$$

Combining this with the alternating binomial coefficients, we obtain

$$\frac{P_3(x_3) - 3P_3(x_2) + 3P_3(x_1) - P_3(x_0)}{h^3} = 6a_0.$$

The constant on the right is the third derivative of the polynomial. In other words: For every polynomial of third degree, the differentiation formula (265) is not just approximately, but exactly correct. In the same way, we could show generally that *the above rule for the numerical differentiation gives exact values for every polynomial of mth degree.* We shall return to this in Chapter 7.

Now we come to a general function $f(x)$ and its given I-values y_0, y_1, y_2, y_3. For $P_3(x)$ we take, in particular, the I-polynomial which assumes these values at the I-places x_0 through x_3. Then the function $f(x) - P_3(x)$ has zeros at these four I-places. According to the theorem of Rolle, a point exists in each one of the intervals (x_0, x_1), (x_1, x_2), (x_2, x_3) at which $f'(x) - P_3'(x)$ vanishes. By applying Rolle's theorem to these three points, we find two points at which $f''(x) - P_3''(x)$ is equal to zero. Finally, a third application of the theorem yields a point ξ in the interval of the I-places at which

$$f'''(\xi) - P_3'''(\xi) = 0, \qquad \text{i.e.,} \qquad f'''(\xi) = P_3'''(\xi).$$

But P_3''' is a constant which, according to our above result, is exactly equal to

$$P_3''' = \frac{y_3 - 3y_2 + 3y_1 - y_0}{h^3}.$$

Hence,

$$f'''(\xi) = \frac{y_3 - 3y_2 + 3y_1 - y_0}{h^3}.$$

The differentiation formula (265) therefore gives the exact value of the third derivative at a point ξ which we cannot identify more closely, but which is certainly contained in the interval of the I-places. We have taken the center of the I-interval for the formulation of the rule of numerical differentiation. This is justified by the fact that ξ is then at most half the length of this interval away from this center. For the general case of the rule we have the same result: The value of the mth derivative obtained by numerical differentiation belongs to an abscissa ξ which is contained in the I-interval. In order to keep the discretization error small we must see to it that this point does not have too much play; i.e., we have to crowd the I-places together as much as possible.

We demonstrate the *numerical error* with the following example. From a table of the exponential function we obtain

$x =$	0.9	1.0	1.1
$f = e^x =$	2.4596	2.7183	3.0042

The second derivative of the function at the point $x = 1$ shall be calculated. Formula (263) with $h = 0.1$ yields

$$f''(1) = 100(3.0042 - 5.4366 + 2.4596) = \underline{2.72}.$$

The exact value of the derivative is the number $e = 2.7183$. We have obtained three correct decimals of this value; however, we note a *loss of significant digits* since the 5-place accuracy of the given I-values was reduced to a 3-place accuracy. For the purpose of diminishing the discretization error, we now make the step length ten times smaller.

$x =$	0.99	1.00	1.01	
$e^x =$	2.6912	2.7183	2.7456	$h = 0.01.$

The same formula yields

$$f''(1) = 10000 \cdot 0.0002 = 2.$$

Here a tremendous loss of significant digits has occurred; we have only obtained one digit, which is not even correct. The automatic computer would interpret this result as 2.0000, continue to calculate with it and finally print out a 5-place result after a large number of computational operations; obviously, not one decimal place of this result would be correct. There is only one remedy for this loss of significant digits. *In order to improve the accuracy of a numerical differentiation not only the I-interval has to be made smaller, but the I-values have to be given with a higher accuracy as well.* If the I-values are the result of a mathematical problem, this is often a very embarrassing condition.

6.11   Other Formulas for Differentiation

In our rule, the mth derivative has been calculated from $(m + 1)$ I-values. It is also possible to use *more I-values* for the calculation of this mth derivative with the intention of improving the accuracy. However, it should thereby be considered that the I-interval will become wider. Appendix II, Table 1, presents some of these formulas. We shall limit ourselves here to deriving only the simplest of these formulas, namely the one in the left-hand upper corner of the table.

Let three I-places x_0, x_1, x_2 and the corresponding I-values y_0, y_1, y_2 of the function $f(x)$ be given; $f'(x_0)$ shall be calculated approximately by means of an equation of the form

$$f'(x_0) \sim p_0 y_0 + p_1 y_1 + p_2 y_2. \tag{266}$$

The coefficients p_0, p_1, p_2 are to be found. To do this, we require that the formula will be exact for three special functions, namely for the first three powers 1, x, x^2 of the variable x. For the sake of simplicity, we choose $x_0 = 0$ and for the step length $h = 1$; this leads to the equations

$$\left. \begin{array}{r} p_0 + p_1 + \ p_2 = 0 \\ p_1 + 2p_2 = 1 \\ p_1 + 4p_2 = 0 \end{array} \right|$$

which have the solution $p_0 = -1.5$, $p_1 = 2$, $p_2 = -0.5$. These are the values in the table.

We take this opportunity to introduce the following important method of deduction called the *superposition principle.* Consider two functions $f_1(x)$, $f_2(x)$ and a linear combination

$$f_3(x) = c_1 f_1(x) + c_2 f_2(x) \tag{267}$$

where c_1, c_2 are given constants. If then the formula (266) is exactly correct for $f_1(x)$ and $f_2(x)$, it is also correct for $f_3(x)$.

Proof. Assume that the I-values of the given functions are

$$\text{for } f_1(x): \quad y_{10}, y_{11}, y_{12}; \quad \text{for } f_2(x): \quad y_{20}, y_{21}, y_{22}.$$

(The first index indicates the number of the function.) Then the I-values of $f_3(x)$ are given by

$$y_{30} = c_1 y_{10} + c_2 y_{20}; \quad y_{31} = c_1 y_{11} + c_2 y_{21};$$

$$y_{32} = c_1 y_{12} + c_2 y_{22}. \tag{268}$$

According to the assumption, we have

$$
\begin{array}{l|l}
f'_1(x_0) = p_0 y_{10} + p_1 y_{11} + p_2 y_{12} & c_1 \\
f'_2(x_0) = p_0 y_{20} + p_1 y_{21} + p_2 y_{22} & c_2
\end{array}
$$

Multiplication of these equations by c_1, c_2 and subsequent addition yields, in view of (268),

$$c_1 f'_1(x_0) + c_2 f'_2(x_0) = p_0 y_{30} + p_1 y_{31} + p_2 y_{32}.$$

This proves the theorem, since according to (267) the left-hand side is the derivative $f'_3(x_0)$.

Basically, the superposition principle is valid here because differentiating is a *linear process* and because (266) contains the interpolating values linearly. It is clear that our theorem also holds if we form linear combinations of more than two functions.

Now we had forced (266) to be exactly correct for the special functions 1, x, x^2. Because of the superposition principle, the formula is then also correct for every linear combination

$$c_1 + c_2 x + c_3 x^2,$$

i.e., for every polynomial of 2nd degree. In particular, the formula is correct for the I-polynomial of 2nd degree which can be passed through the given I-values y_0, y_1, y_2 of $f(x)$. In other words: *The right side of (266) is the derivative of the I-polynomial at x_0* if the above values for p_0, p_1, p_2 are used.

Therefore, this formula could have been derived as well by calculating the I-parabola and differentiating it at the point x_0.

6.2 Numerical Integration

6.21 Trapezoidal Rule

Consider a function $f(x)$ and assume that two I-values y_0, y_1 are given corresponding to the abscissas x_0, x_1 (Fig. 23). From these I-values, the integral

$$\int_{x_0}^{x_1} f(x)\, dx$$

of this function over the I-interval shall be calculated approximately. The value of this integral is equal to the area bounded by the x-axis, the vertical lines through the I-places and the curve $f(x)$. An approximate value for this area is obtained if the area is replaced by the trapezoid which has as its upper boundary the secant A_0A_1 instead of the curve. Its area is equal to $T = (h/2)(y_0 + y_1)$ and we therefore have the approximation formula

$$\int_{x_0}^{x_1} f(x)\, dx \sim T = \frac{h}{2}\,(y_0 + y_1). \tag{269}$$

which is called the *trapezoidal rule*. In order to estimate the discretization error, we use the tangents at the I-points A_0, A_1. The values of the derivative $f'(x)$ at the I-places shall be denoted by

$$y_0' = f'(x_0), \qquad y_1' = f'(x_1).$$

The left tangent intersects the vertical line through the center of the

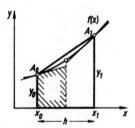

FIG. 23. Trapezoidal rule.

interval at a distance $y_0 + (h/2)y_0'$ from the x-axis. Hence, the area of the shaded trapezoid on the left equals

$$\frac{h}{4}\left[y_0 + \left(y_0 + \frac{h}{2}\,y_0'\right)\right] = \frac{h}{2}\,y_0 + \frac{h^2}{8}\,y_0'.$$

Similarly, the trapezoid formed with the right tangent has the area

$$\frac{h}{2} y_1 - \frac{h^2}{8} y_1'.$$

The sum T^* of these two small trapezoids represents another approximation for the integral

$$T^* = \frac{h}{2} (y_0 + y_1) + \frac{h^2}{8} (y_0' - y_1') = T - \frac{h^2}{8} (y_1' - y_0'). \qquad (270)$$

Now in the case that the curve is convex in the interval (x_0, x_1)—as in Fig. 23 —i.e., that $f''(x) > 0$, then the secant A_0, A_1 is always above the curve and the tangents are always below it. Hence, the value of the integral has to be between the values (269) and (270), and we find for the absolute value of the error of the trapezoidal rule

$$\left| \int_{x_0}^{x_1} f(x)\, dx - T \right| \le \frac{h^2}{8} [f'(x_1) - f'(x_0)].$$

The same holds true if $f''(x) < 0$ in the entire integration interval—in sum, if the curve $f(x)$ has *no inflection point* in the integration interval. In the square brackets we have the change of the derivative $f'(x)$ from one end point of the interval to the other. Hence, the formula is a quantitative form of the qualitative statement that the trapezoidal rule is the more inaccurate, the more strongly the function curves. At the same time, we see that the error decreases whenever h becomes small.

In the evaluation of the trapezoidal rule we must *add* the I-values. Even if the step length h becomes very small, no loss of significant digits need be feared. There is no need to say more about the numerical error.

The formulas (269), (270) certainly give the exact value for the integral if $f(x)$ is a linear function. By linearly combining the values T, T^*, we shall now try to find a formula which even gives the exact integral of a quadratic polynomial. Let p, q be two weights, i.e., two numbers with the sum $= 1$. Then also the formula $(pT + qT^*)$ integrates every linear function correctly. If, in addition, it also gives us correctly the integral of the special function $f(x) = x^2$, then it will also be correct for every quadratic polynomial. This is true since the quadratic polynomial is a linear combination of a linear function and of the function x^2 (superposition principle). Without restricting the generality, we may assume that $x_0 = 0$, $h = 1$. Then we have for the function x^2

$$\int_0^1 x^2\, dx = \tfrac{1}{3}, \qquad T = \tfrac{1}{2}, \qquad T^* = \tfrac{1}{4}.$$

The quantities p, q must be determined in such a way that the equation

$$\tfrac{1}{2}p + \tfrac{1}{4}q = \tfrac{1}{3}$$

is satisfied. Together with $p + q = 1$, this yields $p = \tfrac{1}{3}$, $q = \tfrac{2}{3}$. The formula then is

$$pT + qT^* = T - \frac{h^2}{12}(y_1' - y_0')$$

which provides us with the *improved trapezoidal rule*

$$\int\limits_{x_0}^{x_1} f(x)\, dx \sim \frac{h}{2}(y_0 + y_1) - \frac{h^2}{12}(y_1' - y_0') \qquad (271)$$

y_0, y_1 and y_0', y_1' are the I-values of $f(x)$ and $f'(x)$, respectively.

This formula has the especially welcome property that it is correct even for *polynomials of third degree*. Because of the superposition principle, it is sufficient to verify this for the special function x^3; we leave this to the reader. This property makes (271) a simple and very accurate formula for the numerical integration; however, it is applicable only if the derivative $f'(x)$ is easily accessible.

6.22 Repeated Interval Halving[1]

In order to work out a practical integration method for the integral

$$\int\limits_{a}^{b} f(x)\, dx$$

let us no longer assume that certain I-values of $f(x)$ are given from the very beginning. Rather, we shall assume that we are in a position either to calculate the value of the function for *any* abscissa x or to obtain this value from a table. In the course of the integration process we shall choose the I-places as we find suitable. The given integration interval shall be denoted by $l = (b - a)$; it does not need to be small. In order to save writing effort, we shall normalize this interval to the value $l = 1$ for the following discussion. In addition, we place the lower bound a of the integration at the origin so that we have to evaluate

$$\int\limits_{0}^{1} f(x)\, dx$$

---

[1] Compare also W. Romberg, Vereinfachte numerische Integration. *Kgl. Norske Videnskab. Selskab, Trondheim, Forh.* **28**, No. 7 (1955).

(Fig. 24). The coarsest approximation is given by the trapezoidal value

$$T_1 = \frac{f(0) + f(1)}{2}.$$

To arrive at a better approximation, we halve the interval, determine $f(\frac{1}{2})$ and apply the trapezoidal rule to each one of the subintervals; the left-hand

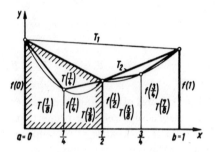

FIG. 24. Repeated halving.

trapezoidal value shall be called $T(\frac{1}{4})$—because $\frac{1}{4}$ is the abscissa of its center point; the right one is called $T(\frac{3}{4})$:

$$T(\tfrac{1}{4}) = \frac{f(0) + f(\frac{1}{2})}{4}, \qquad T(\tfrac{3}{4}) = \frac{f(\frac{1}{2}) + f(1)}{4}.$$

The sum of these two values

$$T_2 = T(\tfrac{1}{4}) + T(\tfrac{3}{4}) = \frac{f(0) + 2f(\frac{1}{2}) + f(1)}{4} \qquad (272)$$

can also be written as

$$T_2 = \tfrac{1}{2}[T_1 + f(\tfrac{1}{2})]. \qquad (273)$$

It is a (possibly better) approximation of the integral. For a second halving step we need $f(\frac{1}{4}), f(\frac{3}{4})$ and the trapezoidal values

$$T(\tfrac{1}{8}) = \frac{f(0) + f(\frac{1}{4})}{8}, \qquad T(\tfrac{3}{8}) = \frac{f(\frac{1}{4}) + f(\frac{1}{2})}{8}$$

$$T(\tfrac{5}{8}) = \frac{f(\frac{1}{2}) + f(\frac{3}{4})}{8}, \qquad T(\tfrac{7}{8}) = \frac{f(\frac{3}{4}) + f(1)}{8}.$$

Their sum

$$T_3 = T(\tfrac{1}{8}) + T(\tfrac{3}{8}) + T(\tfrac{5}{8}) + T(\tfrac{7}{8}) \qquad (274)$$

equals

$$T_3 = \tfrac{1}{8}[f(0) + 2f(\tfrac{1}{4}) + 2f(\tfrac{1}{2}) + 2f(\tfrac{3}{4}) + f(1)]$$

and can also be written in the form

$$T_3 = \frac{1}{2}\left[T_2 + \frac{f(\frac{1}{4}) + f(\frac{3}{4})}{2}\right].$$ (275)

The *generating principle* now becomes clear. In each case, we must add the arithmetic mean of the newly used function values to the previous trapezoidal value. Then the result must be halved. Hence,

$$T_4 = \frac{1}{2}\left[T_3 + \frac{f(\frac{1}{8}) + f(\frac{3}{8}) + f(\frac{5}{8}) + f(\frac{7}{8})}{4}\right]$$ (276)

and so on. In this way, we obtain a sequence of approximations which tends to the value of the integral if $f(x)$ is a continuous function. For example, geometrically we obtain T_4 by replacing the curve $f(x)$ by a *secant polygon* consisting of 8 secants. *Archimedes* used this method for the calculation of an area, namely replacing a curve by inscribed polygons with increasing number of sides. All values T_n are trivially equal to the integral if $f(x)$ is a *linear function*.

Now we shall construct a linear combination of T_1 and T_2 such that the resulting approximation is correct even for a quadratic polynomial $P_2(x)$. We proceed exactly in the same way as the improvement (271) of the trapezoidal rule, i.e., we have to consider only the function $f(x) = x^2$. For it we obtain

$$\int_0^1 x^2\, dx = \tfrac{1}{3}, \qquad T_1 = \tfrac{1}{2}, \qquad T_2 = \tfrac{3}{8}.$$

For the weights p, q needed for generating a formula $pT_1 + qT_2$ we therefore find the equations

$$\tfrac{1}{2}p + \tfrac{3}{8}q = \tfrac{1}{3}, \qquad p + q = 1 \qquad \text{whence} \qquad p = -\tfrac{1}{3}, q = \tfrac{4}{3}.$$

We denote the resulting value by

$$S_1 = \frac{4T_2 - T_1}{3}.$$ (277)

It is not only correct for polynomials of 2nd degree, but also for those of 3rd degree, as we can see by substituting $f(x) = x^3$. The quantity S_1 has the following *geometric meaning*: This value can be calculated as soon as the three function values $f(0), f(\frac{1}{2}), f(1)$ are known. If we pass the I-parabola through these, then S_1 is its integral; this is so because the formula (277) is exact for every $P_2(x)$. This formula therefore means that for the calculation of the integral the function $f(x)$ is replaced by this I-parabola. We now form exactly the same combination following the 2nd halving step in the left subinterval $(0, \frac{1}{2})$. In this interval, the role of T_1, T_2 is assumed by the quantities

$$T(\tfrac{1}{4}) \qquad \text{and} \qquad T(\tfrac{1}{8}) + T(\tfrac{3}{8}).$$

The linear combination therefore yields a value

$$S_1 \text{ (left)} = \frac{4[T(\tfrac{1}{8}) + T(\tfrac{3}{8})] - T(\tfrac{1}{4})}{3}$$

and this is the integral of the I-parabola through $f(0)$, $f(\tfrac{1}{4})$, $f(\tfrac{1}{2})$ taken over the left subinterval.

Similarly, we obtain in the right subinterval

$$S_1 \text{ (right)} = \frac{4[T(\tfrac{5}{8}) + T(\tfrac{7}{8})] - T(\tfrac{3}{4})}{3}$$

as the integral of a second I-parabola determined by the I-values

$$f(\tfrac{1}{2}), \qquad f(\tfrac{3}{4}), \qquad f(1).$$

Under consideration of (274) and (272) the sum of these expressions yields

$$S_2 = S_1 \text{ (left)} + S_1 \text{ (right)} = \frac{4T_3 - T_2}{3}.$$

This expression results if $f(x)$ is replaced by the two mentioned parabolic arcs (they are shown in Fig. 24), and if these are integrated over the entire interval.

Finally,

$$S_3 = \frac{4T_4 - T_3}{3}$$

is the approximate integral which we obtain when replacing $f(x)$ by a curve consisting of 4 parabolic arcs, etc. The values $S_1, S_2, \ldots, S_n, \ldots$ are called *Simpson approximation values*.

This technique of using linear combinations can be continued. We try to determine a linear combination

$$C_1 = pS_1 + qS_2$$

which is not only correct for polynomials of 3rd degree, but also gives the exact integral for those of 4th degree. By substituting $f(x) = x^4$, we find

$$\int_0^1 x^4\,dx = \tfrac{1}{5}, \qquad T_1 = \tfrac{1}{2}, \qquad T_2 = \tfrac{9}{32}, \qquad T_3 = \tfrac{113}{512}$$

$$S_1 = \tfrac{5}{24}, \qquad S_2 = \tfrac{77}{384}$$

and

$$\tfrac{5}{24}p + \tfrac{77}{384}q = \tfrac{1}{5}, \qquad p + q = 1, \qquad p = -\tfrac{1}{15}, \qquad q = \tfrac{16}{15}.$$

This yields the approximation

$$C_1 = \frac{4^2 S_2 - S_1}{4^2 - 1} \tag{278}$$

which is exact even if $f(x)$ is a polynomial of 5th degree; C_1 can be evaluated as soon as we know the five function values $f(0)$, $f(\tfrac{1}{4})$, $f(\tfrac{1}{2})$, $f(\tfrac{3}{4})$, $f(1)$. Hence C_1 is the integral of the I-polynomial of 4th degree determined by these 5 interpolating values. Again the same constructions are made in the subintervals; they lead to the approximations

$$C_2 = \frac{4^2 S_3 - S_2}{4^2 - 1}, \qquad C_3 = \frac{4^2 S_4 - S_3}{4^2 - 1}$$

for the integral of $f(x)$, and so on. These correspond to the replacement of $f(x)$ by arcs of I-polynomials of 4th degree in the separate subintervals. We call them the *Cotes approximations*.

It is essential to study the next linear combination as well. This is a linear combination D_1 of C_1 and C_2 which is correct for polynomials of 6th degree. We find

$$D_1 = \frac{4^3 C_2 - C_1}{4^3 - 1}, \tag{279}$$

and this again is correct even for polynomials of 7th degree; D_1 can be evaluated as soon as the 9 function values $f(0)$, $f(\tfrac{1}{8})$, ..., $f(\tfrac{7}{8})$, $f(1)$ are given. The I-polynomial determined by these 9 interpolating values has the degree 8, and for polynomials of 8th degree the integral-value D_1 is generally no longer correct. From now on, therefore, *the constructed approximations for the integral of $f(x)$ are no longer connected with interpolating polynomials.* We shall see further below that this is actually an advantage. The D sequence, of course, proceeds as follows:

$$D_2 = \frac{4^3 C_3 - C_2}{4^3 - 1}, \qquad D_3 = \frac{4^3 C_4 - C_3}{4^3 - 1}, \;\dots\;.$$

Formulas (277), (278), (279) suggest[1] how the next sequence should be formed:

$$E_1 = \frac{4^4 D_2 - D_1}{4^4 - 1}, \qquad E_2 = \frac{4^4 D_3 - D_2}{4^4 - 1}, \ldots .$$

All these formulas yield exact values for the integral if $f(x)$ is a polynomial of 9th degree. Continuing this, an F-sequence and a G-sequence would appear, and so on. This entire bouquet of approximations for our integral can best be noted down in the following *integration table*:

T_1

$T_2 \quad S_1 = \dfrac{4T_2 - T_1}{3}$

$T_3 \quad S_2 = \dfrac{4T_3 - T_2}{3} \qquad C_1 = \dfrac{4^2 S_2 - S_1}{4^2 - 1}$

$T_4 \quad S_3 = \dfrac{4T_4 - T_3}{3} \qquad C_2 = \dfrac{4^2 S_3 - S_2}{4^2 = 1} \qquad D_1 = \dfrac{4^3 C_2 - C_1}{4^3 - 1}$

$T_5 \quad S_4 = \dfrac{4T_5 - T_4}{3} \qquad C_3 = \dfrac{4^2 S_4 - S_3}{4^2 - 1} \qquad D_2 = \dfrac{4^3 C_3 - C_2}{4^3 - 1} \qquad E_1 = \dfrac{4^4 D_2 - D_1}{4^4 - 1} .$

The table is easily obtained from the first column. Writing (278) in the form

$$C_1 = S_2 + \frac{S_2 - S_1}{15}$$

---

[1] For the advanced reader, we give the following brief derivation of the rule governing the formation of these formulas. Consider any method of numerical integration which is exactly correct for polynomials of degree $(n - 1)$. If now x^n is integrated by this method over any interval of length l, then the resulting value R has an error cl^{n+1} where c is a constant depending neither on the position nor on the length of the integration interval. Now if the interval is halved and the method applied to each one of the subintervals separately, then errors of the magnitude $c(l/2)^{n+1}$ appear on the right as well as on the left. Addition of the contributions from the two subintervals leads to a value R' which has the error

$$2c \left(\frac{l}{2} \right)^{n+1} = \frac{1}{2^n} cl^{n+1}.$$

Finally, if it is desired that the expression $pR + qR'$, formed with the weights p, q, is to integrate the function x^n exactly, then the relations

$$p + q \frac{1}{2^n} = 0, \qquad p + q = 1$$

have to hold, whence

$$p = \frac{-1}{2^n - 1}, \qquad q = \frac{2^n}{2^n - 1}.$$

we can calculate with small corrections. It is possible to show that every column in the table, as well as every downward diagonal, tends to the desired integral value of $f(x)$ provided that $f(x)$ is continuous in the closed integration interval. For a discussion of the speed of convergence and of other properties of the integration method we refer to [36]. All values in the same row use the same I-values. For the practical application, the table must be extended until a cluster of values appears in the lower right which agree in sufficiently many decimal places. For an integration interval of the length $l \neq 1$, one must afterwards still multiply by l.

Preceding these computations a *function table* for the calculation of the T-values has to be set up. We shall arrange that table as follows:

$f(0)$	$f(1)$			M_1	T_1
$f(\frac{1}{2})$				M_2	T_2
$f(\frac{1}{4})$	$f(\frac{3}{4})$			M_3	T_3
$f(\frac{1}{8})$	$f(\frac{3}{8})$	$f(\frac{5}{8})$	$f(\frac{7}{8})$	M_4	T_4

On the right-hand border, the arithmetic mean M of each row is formed. It is then added to the preceding T-value and the result is divided by two. This computation rule was discussed above. In Appendix I, Example 8, the number π has been calculated in this way.

6.23 Supplements to the Integration Theory

Consider Fig. 25 where three equidistant I-values y_0, y_1, y_2 of a given function $f(x)$ have been drawn. The integral of $f(x)$ over the *double interval* (x_0, x_2) is to be calculated approximately. The integration interval has the length $l = x_2 - x_0$. For the trapezoidal values we obtain:

$$T_1 = \frac{l}{2}(y_0 + y_2), \qquad T_2 = \frac{l}{4}(y_0 + 2y_1 + y_2).$$

Hence, from (277) we find the Simpson value

$$S_1 = \frac{l}{6}(y_0 + 4y_1 + y_2).$$

The resulting approximation formula

$$\int_{x_0}^{x_2} f(x)\, dx \sim \frac{l}{6}(y_0 + 4y_1 + y_2) \tag{280}$$

is called the *Simpson rule*. As was shown above, if $f(x)$ is a polynomial of third degree, then the expression on the right yields the value of the integral exactly. This expression, furthermore, is the integral of the interpolating parabola determined by the I-values y_0, y_1, y_2. In general, the Simpson rule is somewhat less accurate than the improved trapezoidal rule (271), but it has the advantage that the derivative $f'(x)$ is not used. The coefficients 1, 4, 1 in (280) are called *integration weights*; the denominator 6 must of course be equal to their sum in order for the formula to give the right value for the constant $f(x) \equiv 1$.

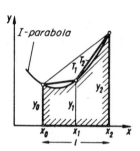

FIG. 25.   Simpson rule; S_1 = shaded area.

Now let 5 equidistant I-places x_0, ... , x_4 and the corresponding I-values y_0, ... , y_4 be given and let $l = x_4 - x_0$ be the length of the I-interval. For the integration over this interval, we can apply the Simpson rule to the subintervals (x_0, x_2) and (x_2, x_4) separately, and then add the results. In this way we find:

$$\int_{x_0}^{x_4} f(x)dx \sim \frac{l}{12} (y_0 + 4y_1 + 2y_2 + 4y_3 + y_4)$$

which, incidentally, has to be the value S_2, according to our theory. One proceeds in the same way if a large *odd* number of interpolating values is available. The inside I-values then get the weights 4, 2, alternatively and the two I-values on the ends of the interval acquire the weight 1.

If we apply the *improved trapezoidal rule* (271) in the same manner to a sequence of consecutive subintervals, we happily find that the values of the derivative at the inside I-places drop out, so that $f'(x)$ needs to be known only at the two end points of the total integration interval.

We have dealt with these integration methods only summarily since they do not give as much information as the complete integration table.

Appendix II, Table 2, contains the integration weights for some additional formulas for approximate integration. With the help of the principle of superposition, they were derived as follows: If $(n + 1)$ I-points are given, the formula looked for is first set up with unknown integration weights and then it is demanded that this formula yield the correct integral for all the powers $1, x, x^2, \ldots, x^n$. The result will then be automatically correct for the I-polynomial.

Among these formulas those which integrate the I-polynomial over the entire I-interval are called *Cotes formulas*. For the special case of five I-points the Cotes formula yields the value C_1 of our integration table. For a number of at most eight I-points all the weights in the Cotes formulas are positive. However, for the formula using nine I-points, *negative weights* occur for the first time. Such a formula is very suspect. In fact, the integral is by definition the limit of a *sum of ordinates* inside the integration interval. There is no reason why certain of these ordinates should be taken negatively.

It is exactly at this critical point that the method of our integration table departs from the Cotes procedure, since for the first time the value D_1 of (279) no longer has anything to do with the I-polynomial, even though this value, too, is determined by nine function values. When D_1 is represented explicitly as linear combination of these nine values we find that all integration weights are positive.

6.24 Periodic Functions

Let $f(x)$ be continuous and periodic and let the length of the period be $= 1$. The integral J over this period is to be calculated approximately. It is well known that J does not depend on the position of the periodicity interval; any interval of the length 1 can be taken. To be more specific, we shall subdivide this interval into 4 equal parts for our numerical integration, so that 5 I-places appear. Because of the periodicity the first and last of the I-values are equal and we can therefore combine them in the integration formula. In this way a linear combination of the first 4 I-values results; we shall denote its coefficients by p_0, p_1, p_2, p_3. If we place the left end point of the interval at the origin this formula gives the value

$$J \sim p_0 f(0) + p_1 f(\tfrac{1}{4}) + p_2 f(\tfrac{1}{2}) + p_3 f(\tfrac{3}{4}).$$

However, if we place the left end point at the point $x = \tfrac{1}{4}$, then the value

$$J \sim p_0 f(\tfrac{1}{4}) + p_1 f(\tfrac{1}{2}) + p_2 f(\tfrac{3}{4}) + p_3 f(0).$$

results. In the last term we have made use of the periodicity. Since the left-hand sides are equal in both expressions it is reasonable to require that the approximations on the right-hand side agree as well. For arbitrary periodic functions this is possible only if, one by one, the following relations hold:

$$p_1 = p_0, \qquad p_2 = p_1, \qquad p_3 = p_2, \qquad p_0 = p_3,$$

i.e., if the *integration weights* are equal. The formula therefore must read:

$$J \sim \tfrac{1}{4}[f(0) + f(\tfrac{1}{4}) + f(\tfrac{1}{2}) + f(\tfrac{3}{4})].$$

The denominator 4 is again obtained from the consideration that the formula has to yield the correct integration value for a constant function. Hence, a *periodic function* is *integrated* numerically over its periodicity interval by subdividing this interval into n equal parts and calculating the *arithmetic mean* of the n ordinates corresponding to the left end points of each subinterval. If the periodicity interval has the length l, the result must still be multiplied by l.

These considerations become even clearer if we regard the periodic function as a function on the circumference of a circle.

6.3 Differential Equations of the First Order

The discretization of an ordinary differential equation will be explained with the help of the following example of a *linear* differential equation. Let $y(x)$ be an unknown function satisfying the relation

$$y' = -xy \tag{281}$$

where y' denotes the derivative of y with respect to x. In addition, at the point x_0 the *initial condition*

$$y(x_0) = y_0 \tag{282}$$

has been given; here y_0 is any prescribed value. It is desired to calculate $y(x)$ for additional values of the argument x, and, in particular, first at a point x_1 in the neighborhood of x_0. The quantity $h = x_1 - x_0$ is therefore to be the first *step* that we shall make (Fig. 23).

The basic idea for solution of the differential equation (281) consists in integrating both sides of the equation over the interval from x_0 to x_1:

$$\int_{x_0}^{x_1} y' \, dx = - \int_{x_0}^{x_1} xy \, dx.$$

The integral on the left can be evaluated immediately and yields $y(x_1) - y(x_0)$:

$$y(x_1) = y(x_0) - \int_{x_0}^{x_1} xy \, dx. \tag{283}$$

It is different with the remaining integral. Since the behavior of the function $y(x)$ in the integration interval is totally unknown, we must resort to an approximate numerical integration. We shall carry this out with the help of the trapezoidal rule (269). In this connection it must be noted that the integrand is the function $xy(x)$; the trapezoidal rule therefore yields

$$\int_{x_0}^{x_1} xy\, dx \sim \frac{h}{2}\left[x_0 y(x_0) + x_1 y(x_1)\right]$$

whence (283) becomes

$$y(x_1) \sim y(x_0) - \frac{h}{2}\left[x_0 y(x_0) + x_1 y(x_1)\right].$$

In this approximate equation $y(x_0) = y_0$ is given by the initial condition, and, accordingly, the new function value $y(x_1)$ is the only unknown. It is better to write an exact equation, but then we should remember that we obtain only an approximate value of $y(x_1)$; this value will be denoted by y_1:

$$y_1 = y_0 - \frac{h}{2}\left(x_0 y_0 + x_1 y_1\right). \tag{284}$$

As a result of the discretization, the *linear differential equation* for $y(x)$ has become an *ordinary linear equation* for the determination of the approximate value y_1 of $y(x_1)$. Its solution yields

$$y_1 = \frac{2 - hx_0}{2 + hx_1}\, y_0. \tag{285}$$

Naturally, the approximation value y_1 will be the more accurate the smaller the step h has been chosen.

Now follows a second integration step which uses the already known value y_1 as initial value, and which leads to a point x_2. We shall assume that the second step $(x_2 - x_1)$ also has the length h, so that the approximated value y_2 of $y(x_2)$ is obtained:

$$y_2 = \frac{2 - hx_1}{2 + hx_2}\, y_1.$$

Since y_1 was already inaccurate, y_2 will approximate the value $y(x_2)$ with increased inaccuracy. In other words, we have to reckon with a magnification of the discretization error in the course of our integration procedure. For this reason, one should not leave the step constant in the practical application, but should decrease it, for example, when coming into areas where $y(x)$ varies more strongly.

The continuation of the integration procedure now results in the general *recursion formula*

$$y_{n+1} = \frac{2 - hx_n}{2 + hx_{n+1}} y_n \qquad (286)$$

for the approximate calculation of a table of the unknown function $y(x)$ with a constant step h.

Numerical Example.

$$x_0 = 0, \qquad y_0 = 1, \qquad h = 0.1;$$

$$y_1 = \frac{2}{2 + 0.01} = 0.995, \qquad y_2 = \frac{2 - 0.01}{2 + 0.01} 0.995 = 0.980,$$

$x =$	0	0.1	0.2	0.3	0.4	0.5
$y =$	1	0.995	0.980	0.956	0.923	0.882

$$(287)$$

The exact solution of the differential equation is

$$y = e^{-x^2/2};$$

the tabulated function values are accurate to three decimal places.

The main difficulty with the numerical integration of differential equations is the choice of an *advantageous step length* and its modification during the integration process. While it is true that a small h keeps the discretization error down, it encourages the accumulation of numerical errors because of the large number of integration steps. Information about the discretization error can be obtained if, for the evaluation of the integral in (283), we use a more accurate rule for the numerical integration. The improved trapezoidal rule (271) suggests itself here. In order to be able to evaluate the correction term in that formula we need the derivative of the integrand $xy(x)$:

$$\frac{d}{dx}(xy) = xy' + y.$$

By taking y' from the given differential equation (281), we find

$$\frac{d}{dx}(xy) = y(1 - x^2)$$

and the improved trapezoidal rule yields for our integral

$$\int_{x_0}^{x_1} xy\, dx \sim \frac{h}{2}\left[x_0 y(x_0) + x_1 y(x_1)\right] - \frac{h^2}{12}\left[y(x_1)(1 - x_1^2) - y(x_0)(1 - x_0^2)\right]$$

whence (283) assumes the form

$$y(x_1) \sim y_0 - \frac{h}{2}[x_0 y_0 + x_1 y(x_1)] + \frac{h^2}{12}[y(x_1)(1 - x_1^2) - y_0(1 - x_0^2)].$$

If, in further approximation, we replace $y(x_1)$ on the right by y_1, we find because of (284)

$$y(x_1) \sim y_1 + \frac{h^2}{12}[y_1(1 - x_1^2) - y_0(1 - x_0^2)] \tag{288}$$

where y_1 must be calculated according to (285). Here h must be chosen so small that the correction to be added to y_1 remains below a given tolerance. In our above numerical example this correction has the value

$$\frac{0.01}{12}[0.995 \cdot 0.99 - 1] = -0.000012. \tag{289}$$

For the desired three-decimal-place accuracy of the function values the desired step length is therefore almost too small.

At the same time, the formula (288) presents us with a more accurate numerical procedure for the solution of our differential equation. By calculating with more places we find

$$y_1 = \frac{2}{2 + 0.01} = 0.995025$$

and by addition of the correction (289)

$$y(x_1) = 0.995013.$$

This agrees with the value of the exact solution of the differential equation up to one unit in the last decimal place.

Basically, this correction method can be applied to every differential equation, regardless of whether it is linear or not. The only condition is that the derivatives appearing in the correction term of the improved trapezoidal rule (271) can be calculated without too much trouble. The method can then indeed be recommended highly because of its high accuracy and its numerical stability.

Very detailed discussions of the discretization error and the numerical error for different integration procedures for ordinary differential equations can be found in [20]. In practical applications we will often have to be satisfied with repeating the integration with smaller step lengths until the function values change only within the prescribed tolerances.

6.31   Nonlinear Differential Equations of the First Order

The general differential equation of the first order has the form

$$y' = f(x, y). \tag{290}$$

Here the right-hand side is a given function of the two variables, x, y. In addition we have an initial condition at the point x_0:

$$y(x_0) = y_0 \tag{291}$$

where y_0 is given. As in our special example, we choose a first step h which leads to the point $x_1 = x_0 + h$. Then we integrate the differential equation over the interval (x_0, x_1):

$$y(x_1) = y(x_0) + \int_{x_0}^{x_1} f(x, y)\, dx.$$

Using the trapezoidal rule we find for the approximate value y_1 of $y(x_1)$ the equation

$$y_1 = y_0 + \frac{h}{2}[f(x_0, y_0) + f(x_1, y_1)]. \tag{292}$$

Here the unknown y_1 appears on the right as argument of the function f. If the given *differential equation* (290) is *nonlinear*, i.e., if f—as function of y—is not a linear function, then discretization has resulted in a *nonlinear equation* (292) for y_1. This equation is best solved by an iterative method (4.2): on the right an approximate value is substituted for y_1, then the left-hand side will be an improved approximation, etc.

A first approximate value for y_1 is easily obtained by the following consideration: Because of the initial condition we know at the initial point x_0 the value y_0 of the function sought, and because of the differential equation we also know the value

$$y'_0 = f(x_0, y_0)$$

of its derivative. In Fig. 26 a point A is therefore known, as well as the tangent of the curve $y(x)$ at this point. In our interval (x_0, x_1) we now replace the curve by its tangent; in other words, we *linearize*. This yields the approximation

$$y_1^* = y_0 + hy'_0 = y_0 + hf(x_0, y_0) \tag{293}$$

for y_1. This formula is called the *predictor*, since it predicts a first value of y_1. By inserting y_1^* on the right of (292), we obtain the improved approximation

$$y_1 = y_0 + \frac{h}{2}[f(x_0, y_0) + f(x_1, y_1^*)]. \tag{294}$$

With this we shall stop the iteration. (Actually, we should insert this value again on the right of (292) and so on.) This formula (294) is called the *corrector*, since it corrects the approximation y_1^*. If, for the computation of $y(x_1)$, we content ourselves with the predictor y_1^*, this is called the *method of Euler*. By adding the use of the corrector, the improved Euler method results; it is also called the *method of Heun*. The discretization error of this method consists of two parts; namely, first, of the error caused by integration with the help of the trapezoidal rule, and second, of the error caused

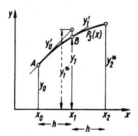

FIG. 26.   Methods of Heun and Runge-Kutta.

by premature interruption of the iteration. In the next paragraph we shall summarize the computational procedure thereby using the abbreviations

$$f_0 = f(x_0, y_0), \qquad f_1^* = f(x_1, y_1^*). \tag{295}$$

6.32   Method of Heun

This is a method for the solution of $y' = f(x, y)$ under the initial condition $y(x_0) = y_0$. Choose a step h, $x_1 = x_0 + h$:

$$\boxed{\text{Predictor} \quad y_1^* = y_0 + hf_0, \qquad \text{Corrector} \quad y_1 = y_0 + \frac{h}{2}(f_0 + f_1^*)} \tag{296}$$

Using x_1, y_1 as initial data, the second integration step follows, and so on.

Example.   For the differential equation (281), we have

$$f(x, y) = -xy.$$

With the special assumption of example (287), it follows, that

$$f_0 = 0, \quad y_1^* = 1, \quad f_1^* = -0.1, \quad y_1 = 1 + 0.05\,(-0.1) = 0.995 \tag{297}$$

which agrees with the table (287).

While the Heun method is crude, it is also a simple and very stable method for the solution of differential equations.

6.33   Methods of Runge-Kutta

The idea suggests itself of improving our methods by using, instead of the trapezoidal rule, a finer integration rule, as for example the Simpson rule. We shall begin with the following situation (Fig. 26): Assume that besides the initial value $y_0 = y(x_0)$ the value $y_1 = y(x_1)$ at the end of the first step is already known. We now add a second step of the same length h which leads to x_2; then we integrate the differential equation (290) over the double interval from x_0 to x_2, and obtain

$$y(x_2) = y(x_0) + \int_{x_0}^{x_2} f(x, y)\, dx. \tag{298}$$

In the integrand $f(x, y)$ we must, of course, replace y by the desired function $y(x)$. The values of the integrand at the points x_0, x_1, x_2 therefore are

$$f(x_0, y_0), \qquad f(x_1, y_1), \qquad f(x_2, y(x_2)),$$

respectively.

The integration interval has the length $l = 2h$ and hence the Simpson rule (280) yields

$$\int_{x_0}^{x_2} f(x, y)\, dx \sim \frac{h}{3}\, [f(x_0, y_0) + 4f(x_1, y_1) + f(x_2, y(x_2))];$$

(298) then becomes

$$y(x_2) \sim y(x_0) + \frac{h}{3}\, [f(x_0, y_0) + 4f(x_1, y_1) + f(x_2, y(x_2))].$$

Instead of this approximate equation, we can again write an exact one by introducing in place of $y(x_2)$ an approximation y_2 of this value:

$$y_2 = y_0 + \frac{h}{3}\, [f(x_0, y_0) + 4f(x_1, y_1) + f(x_2, y_2)]. \tag{299}$$

This is, in general, a nonlinear equation for y_2; therefore we again need a predictor in order to find a first approximation y_2^* of y_2. It is logical to construct this predictor in the same way as was done for the Heun method. At the points x_0, x_1 we know not only the values y_0, y_1 of the unknown function, but—because of the differential equation—also the values

$$y_0' = f(x_0, y_0), \qquad y_1' = f(x_1, y_1) \tag{300}$$

of its derivative. In Fig. 26, therefore, we know two points A, B on the sought curve $y(x)$, together with the corresponding tangents. Hence, we can approximate $y(x)$ by a polynomial of third degree $P_3(x)$, which passes through the same points, has the same tangents, and therefore satisfies the conditions

$$P_3(x_0) = y_0, \quad P_3'(x_0) = y_0', \quad P_3(x_1) = y_1, \quad P_3'(x_1) = y_1'. \quad (301)$$

In order not to interrupt our discussion by some incidental considerations, we simply give here the equation of this cubic curve and verify afterwards that it does indeed satisfy the stated conditions. With the abbreviations

$$p = x - x_0, \qquad q = x - x_1$$

this curve has the form

$$P_3(x) = \frac{1}{h^3}[q^2(3p - q)y_0 - p^2(3q - p)y_1] + \frac{1}{h^2}[pq^2 y_0' + p^2 q y_1']. \quad (302)$$

Differentiation immediately yields

$$P_3'(x) = \frac{6pq}{h^3}(y_0 - y_1) + \frac{1}{h^2}[q(q + 2p)y_0' + p(p + 2q)y_1'].$$

At the point x_0 we have $p = 0$, $q = -h$, and hence these formulas indeed give $P_3(x_0) = y_0$ and $P_3'(x_0) = y_0'$. Similarly, we verify that the two remaining conditions (301) are satisfied. At the point x_2 we have $p = 2h$, $q = h$, and therefore as value of the polynomial

$$y_2^* = P_3(x_2) = 5y_0 - 4y_1 + 2hy_0' + 4hy_1'.$$

This is our predictor formula. Substitution in (299) produces the corrector formula

$$y_2 = y_0 + \frac{h}{3}[f(x_0, y_0) + 4f(x_1, y_1) + f(x_2, y_2^*)]$$

and with this we again stop the iterative solution of (299).

In order to arrange the formulas conveniently for practical application, we introduce the abbreviations

$$f_0 = f(x_0, y_0), \qquad f_1 = f(x_1, y_1), \qquad f_2^* = f(x_2, y_2^*). \quad (303)$$

Then, because of (300), the predictor has the form

$$y_2^* = 5y_0 - 4y_1 + 2hf_0 + 4hf_1 \quad (304)$$

and the corrector

$$y_2 = y_0 + \frac{h}{3}(f_0 + 4f_1 + f_2^*).$$ (304a)

But our method still suffers from the disadvantage that the value y_1 of the unknown function at the point x_1 was assumed to be known. We can remedy this by calculating y_1 with the help of the formula (296) of Heun. Substituting this value of y_1 in (304), we find for the predictor the simple and final form

$$y_2^* = y_0 - 2hf_1^* + 4hf_1.$$

Combination of the Heun formulas with our new ones produces the following set of computational rules:

$$y_1^* = y_0 + hf_0, \qquad y_1 = y_0 + \frac{h}{2}(f_0 + f_1^*)$$

$$y_2^* = y_0 - 2hf_1^* + 4hf_1, \qquad y_2 = y_0 + \frac{h}{3}(f_0 + 4f_1 + f_2^*)$$

where the abbreviations (295) and (303) should be observed. In the general literature these rules usually appear in a somewhat different form. To establish the connection here, we introduce the following quantities:

$$k_1 = 2hf_0 = 2hf(x_0, y_0)$$

$$k_2 = 2hf_1^* = 2hf(x_1, y_1^*) = 2hf\left(x_0 + h, y_0 + \frac{k_1}{2}\right)$$

$$k_3 = 2hf_1 = 2hf(x_1, y_1) = 2hf\left(x_0 + h, y_0 + \frac{k_1}{4} + \frac{k_2}{4}\right)$$

$$k_4 = 2hf_2^* = 2hf(x_2, y_2^*) = 2hf(x_0 + 2h, y_0 - k_2 + 2k_3).$$

Then

$$y_2 = y_0 + \tfrac{1}{6}(k_1 + 4k_3 + k_4).$$

After this compact set of formulas has been derived, it is no longer necessary to hold on to the idea that two integration steps have been executed—namely the first one in the interval (x_0, x_1) with the help of Heun's method, and the second one in the interval (x_1, x_2) with the help of Simpson's rule. Instead, we combine everything into one integration step which has the step length h and leads from x_0 to x_1. This change merely results in having to replace $2h$ by h and y_2 by y_1 in the above formulas. We summarize this as follows:

I. First Runge-Kutta Method

Consider the differential equation $y' = f(x, y)$ together with the initial condition $y(x_0) = y_0$. Choose a step h, which leads from x_0 to $x_1 = x_0 + h$. Perform the computations according to the following system of formulas:

$$\boxed{\begin{aligned} k_1 &= hf(x_0, y_0) \\[2mm] k_2 &= hf\left(x_0 + \frac{h}{2}, y_0 + \frac{k_1}{2}\right) \\[2mm] k_3 &= hf\left(x_0 + \frac{h}{2}, y_0 + \frac{k_1}{4} + \frac{k_2}{4}\right) \\[2mm] k_4 &= hf(x_0 + h, y_0 - k_2 + 2k_3) \\[2mm] y_1 &= y_0 + \tfrac{1}{6}(k_1 + 4k_3 + k_4). \end{aligned}} \tag{305}$$

Then y_1 is an approximation for the function value $y(x_1)$. Using the initial data a second integration step follows (it may have another step length), and so on.[1]

Example. For the differential equation (281) together with the special assumptions $x_0 = 0$, $y_0 = 1$, $h = 0.1$ of the table (287), we find

$$k_1 = 0, \quad k_2 = -0.005, \quad k_3 = -0.00499375, \quad k_4 = -0.009950125,$$

$$y_1 = \underline{\underline{0.995012479.}}$$

This agrees in all places with the value of the exact solution for the differential equation.

For practical applications a system of formulas has become standard which is more convenient than (305) and provides about the same integration accuracy. It reads as follows:

---

[1] One could conceivably get the idea here of continuing the computations according to the formulas (304), (304a), and in the case of a linear differential equation, one might even consider solving the Simpson formula

$$y_2 = y_0 + \frac{h}{3}(f_0 + 4f_1 + f_2)$$

for the value y_2 (which also appears in f_2). However, because of their tendency toward numerical instability, such methods cannot be recommended.

2. Second Method of Runge-Kutta

$$
\begin{aligned}
k_1 &= hf(x_0, y_0) \\[6pt]
k_2 &= hf\left(x_0 + \frac{h}{2}, y_0 + \frac{k_1}{2}\right) \\[6pt]
k_3 &= hf\left(x_0 + \frac{h}{2}, y_0 + \frac{k_2}{2}\right) \\[6pt]
k_4 &= hf(x_0 + h, y_0 + k_3) \\[6pt]
y_1 &= y_0 + \tfrac{1}{6}(k_1 + 2k_2 + 2k_3 + k_4)
\end{aligned}
\tag{306}
$$

The derivation can no longer be based on the Simpson rule; in this connection we refer to [20] and [21]. Furthermore, our derivation is rather to be regarded as heuristic; it provides no justification for the fact that for a small step length the Runge-Kutta formula shows a very small discretization error.

It should also be noted that in the case of a linear differential equation both variants (305) and (306) give exactly the same value for y_1. The reader can easily verify this for himself.

Finally, it must be stressed that the calculation of the *integral function* of a given *function $f(x)$* appears as a special case here; in fact, this problem is equivalent to solving the special differential equation

$$y' = f(x).$$

For this special case the Heun method becomes the ordinary trapezoidal rule and the two Runge-Kutta methods reduce to the Simpson rule.

6.34　Method of Adams

In principle, all rules for the numerical integration can also be used for the solution of differential equations. For example, let us assume that for the solution $y(x)$ of the differential equation

$$y' = f(x, y)$$

three values

$$y_0 = y(x_0), \qquad y_1 = y(x_1), \qquad y_2 = y(x_2)$$

are already known, whereby the abscissas x_0, x_1, x_2 follow each other in equal steps of the length h. The next value $y(x_3)$ at the point $x_3 = x_2 + h$ is to be found. Integration over the interval (x_2, x_3) yields

$$y(x_3) = y(x_2) + \int_{x_2}^{x_3} f(x, y) \, dx. \tag{306a}$$

For the numerical integration the rule of Appendix II, Table 2, can be used (integration over an overhanging interval, three I-places). It gives us

$$y(x_3) \sim y_2 + \frac{h}{12} \left[5f(x_0, y_0) - 16f(x_1, y_1) + 23f(x_2, y_2) \right]. \tag{307}$$

This is not an equation but an explicit formula for $y(x_3)$. However, this convenient method for the solution of a differential equation has two disadvantages: First, it can be applied only if some values of the function sought have already been obtained from other sources; second, it assumes for its application a constant step length while it is often necessary to shorten the step length when the integration enters into regions where the course of $y(x)$ is subject to quick changes.

After $y_3 = y(x_3)$ has been obtained from (307), we can use the same formula and the values y_1, y_2, y_3 to calculate the next value $y(x_4)$, and so on.

Example.   For the differential equation (281) we find from table (287)

$$f(x_0, y_0) = -x_0 y_0 = 0, \qquad f(x_1, y_1) = -0.1 \cdot 0.995 = -0.0995$$

$$f(x_2, y_2) = -0.2 \cdot 0.980 = -0.1960.$$

With $h = 0.1$, (307) then gives

$$y(x_3) = 0.980 - \frac{0.1}{12} \, 2.916 = 0.956$$

and this agrees with the next value in table (287).

A variant of the Adams Method should also be mentioned here. It consists in using the appropriate rule of Applendix II, Table 2, for the integration over the *last interval* to evaluate the integral (306a). In this way we obtain

$$y(x_3) = y(x_2) + \frac{h}{24} \left[f(x_0, y_0) - 5f(x_1, y_1) + 19f(x_2, y_2) + 9f(x_3, y_3) \right]. \tag{307a}$$

Here the unknown $y(x_3) = y_3$ again appears on the right in the last term and y must therefore be determined iteratively from this equation. As a predictor formula we can use (307); (307a) then is the corresponding corrector formula.

We can of course use more than three y-values as initial values for the direct method (307a) as well as for the more accurate indirect method (307a).

The coefficients in formulas (307), (307a)—and even more those in formulas with more than three I-places—make us fear the loss of decimals. To counteract this possibility, it is better to use the formulas of the difference calculus for the numerical integration (compare end of 7.12).

6.4 Systems of Differential Equations of the First Order

As an introductory example we make use of the electrical oscillator shown in Fig. 27, in which a capacitor with capacitance C and an impedance coil with the inductance L are connected in series. If this oscillator is left to

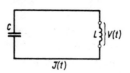

FIG. 27. Oscillator.

itself, the current will be a function $J(t)$ of the time and so will be the voltage $V(t)$ at the coil. These two functions satisfy a system of *linear* differential equations

$$\frac{dJ}{dt} = \frac{1}{L}\,V(t), \qquad \frac{dV}{dt} = -\frac{1}{C}\,J(t).$$ (308)

For both unknown functions, initial conditions of the form

$$J(0) = J_0, \qquad V(0) = V_0$$

must have been given, where J_0, V_0 are prescribed values. Such systems of differential equations are solved according to exactly the same numerical method as single differential equations. A small step h of the time is chosen and both differential equations are integrated over it:

$$\int_0^h \frac{dJ}{dt}\,dt = \frac{1}{L} \int_0^h V(t)\,dt, \qquad \int_0^h \frac{dV}{dt}\,dt = -\frac{1}{C} \int_0^h J(t)\,dt.$$

The integrals on the left can again be calculated exactly; this leads to the equations

$$J(h) - J_0 = \frac{1}{L} \int_0^h V(t)\,dt, \qquad V(h) - V_0 = -\frac{1}{C} \int_0^h J(t)\,dt.$$

We integrate the remaining integrals by means of the trapezoidal rule, thereby obtaining the equations

$$J_1 - J_0 = \frac{h}{2L}(V_0 + V_1), \qquad V_1 - V_0 = -\frac{h}{2C}(J_0 + J_1) \qquad (309)$$

for the approximations J_1, V_1 of the unknown function values $J(h)$, $V(h)$. Here J_0, V_0 are known. The discretization therefore has changed the system (308) of *linear* differential equations to an ordinary *linear* system of equations for the unknowns J_1, V_1.

$$J_1 = \frac{\left(1 - \dfrac{h^2}{4CL}\right)J_0 + \dfrac{h}{L}V_0}{1 + \dfrac{h^2}{4CL}}, \qquad V_1 = \frac{-\dfrac{h}{C}J_0 + \left(1 - \dfrac{h^2}{4CL}\right)V_0}{1 + \dfrac{h^2}{4CL}}$$

is the solution of this system of equations. This provides us with recursion formulas which permit the computation of J_1 and V_1 from J_0, V_0. Increasing all indices by 1, we can also apply them to the second interval and so obtain approximations J_2, V_2 for $J(2h)$, $V(2h)$, and so on.

The other methods presented in 6.3 for the solution of a single differential equation can easily be extended to systems as well. For example, consider the system of coupled differential equations

$$y' = f(x, y, z), \qquad z' = g(x, y, z)$$

for the determination of two functions $y(x)$ and $z(x)$. Let

$$y(x_0) = y_0, \qquad z(x_0) = z_0$$

be the initial conditions. As a sample we give the formulas for the second method of Runge-Kutta in the form of the usual computational scheme

x	y	z	k	l	
x_0	y_0	z_0	k_1	l_1	
$x_0 + \dfrac{h}{2}$	$y_0 + \dfrac{k_1}{2}$	$z_0 + \dfrac{l_1}{2}$	k_2	l_2	K
$x_0 + \dfrac{h}{2}$	$y_0 + \dfrac{k_2}{2}$	$z_0 + \dfrac{l_2}{2}$	k_3	l_3	L
$x_0 + h$	$y_0 + k_3$	$z_0 + l_3$	k_4	l_4	
$x_0 + h$	$y_1 = y_0 + K$	$z_1 = z_0 + L$			

$$(310)$$

Every k (or l) value is obtained by evaluating the function f (or g, respectively) at the arguments written in the first part of the same row, and by multiplying the result by h. Moreover,

$$K = \tfrac{1}{6}(k_1 + 2k_2 + 2k_3 + k_4), \qquad L = \tfrac{1}{6}(l_1 + 2l_2 + 2l_3 + l_4).$$

We will have an opportunity later on to add a computational example to this table.

6.41   Differential Equations of Higher Order

Such differential equations are usually solved by reducing them to systems of first order. For example, let us assume that the one-dimensional wave equation

$$y'' + p(x)y = 0$$

has to be solved for a function $y(x)$, subject to the initial conditions

$$y(0) = y_0, \qquad y'(0) = y_0'.$$

Here the function $p(x)$ is *positive* for $x \geq 0$. We consider the derivative $y'(x)$ as a second unknown function $z(x)$:

$$y' = z. \tag{311}$$

The given differential equation can then be written in the form

$$z' = -p(x)y. \tag{312}$$

Together, (311) and (312) form a system of differential equations of first order with the initial conditions

$$y(0) = y_0, \qquad z(0) = z_0 = y_0'.$$

In the case of such *transient problems* the step length h for the numerical integration is chosen in accordance with the following points of view. For small x, $y(x)$ will not deviate too much from the solution $\eta(x)$ of the differential equation

$$\eta'' + p(0)\eta = 0$$

which is obtained by approximating $p(x)$ by a constant. The solution of this differential equation is a harmonic oscillation with the period $2\pi/\sqrt{p(0)}$. In order to reproduce the oscillation correctly, one should compute at least 12 function values per period in the course of the numerical integration. Hence,

$$h < \frac{1}{2\sqrt{p(0)}}$$

provides us with a rule of thumb for choosing the initial step length. Depending on the desired accuracy, a coarser integration method (trapezoidal rule) will suffice, or else a finer method (Runge-Kutta) must be used.

Numerical methods have also been worked out which integrate a differential equation of the second or higher degree directly, without using the reduction to a system of the first order [21]. However, detailed studies, [20], [22], show the following: With respect to the discretization error, these latter methods may be more accurate during the first integration step than the reduction to a system while requiring a similar amount of computation—but this no longer holds true for many integration steps.

The general differential equation of the second order has the form:

$$y'' = f(x, y, y').$$

(313)

In celestial mechanics and also in other areas of applied mathematics, one often runs into differential equations of the special form

$$y'' = f(x, y)$$

where the first derivative is missing on the right-hand side. In this special case there are methods which avoid the reduction to a system, thereby providing small advantages [20]. However, special measures must then be taken to prevent a dangerous accumulation of round-off errors.

6.42 Singularities

We are frequently faced with the problem of having to integrate a differential equation (313) starting with an initial point x_0 at which the function $f(x, y, y')$ is singular. For example, it can happen that at the point x_0, f becomes infinite or has one of the other singularities described and classified in complex function theory. Such an occurrence is perfectly normal; for example, it takes place regularly when *polar coordinates* are used for the solution of a problem. The polar coordinate system is indeed singular at the origin because the polar angle is undetermined at that point. The discussion which follows will illustrate this more specifically.

The partial differential equation

$$\frac{\partial^2 u}{\partial x^2} + \frac{\partial^2 u}{\partial y^2} + \omega^2 u = 0$$

plays an important role in the theory of wave propagation. Here $u(x, y)$ is an unknown function of the two rectangular coordinates x, y, and ω is a given frequency. After transformation of this equation into polar coordinates r, φ, it assumes the form

$$\frac{\partial^2 u}{\partial r^2} + \frac{1}{r} \frac{\partial u}{\partial r} + \frac{1}{r^2} \frac{\partial^2 u}{\partial \varphi^2} + \omega^2 u = 0.$$

For simplicity's sake we assume that u depends only on the radius r, thereby obtaining the ordinary differential equation

$$\frac{d^2u}{dr^2} + \frac{1}{r}\frac{du}{dr} + \omega^2 u = 0.$$

By means of the substitution $\rho = \omega r$ this can be simplified to

$$\frac{d^2u}{d\rho^2} + \frac{1}{\rho}\frac{du}{d\rho} + u = 0.$$

If, instead of u, ρ, we again write y, x, we arrive at the special *Bessel differential equation*

$$y'' = -\frac{1}{x}y' - y. \tag{314}$$

This is a linear differential equation and a special case of (313). At the origin $x_0 = 0$ it has a singularity because there the coefficient of y' becomes infinite.

In such a case of singular behavior at the initial point x_0, the only thing we can do is to study the nature of the singularity with the methods of real and complex analysis [23]. These investigations end with a *series expansion* which permits the computation of the function y and of its derivative at a point x_1 in the neighborhood of x_0. Then the numerical integration can be started at this point x_1 because now $y(x_1)$ and $y'(x_1)$ have become known as initial values. It should be mentioned that this *method of series expansion* can also be used in a regular initial point x_0. In that case, the series is an ordinary *Taylor series* and can be used, for example, as a means to start the Adams method (6.34). From this series y and y' are calculated at the points $x_0 + h$, $x_0 + 2h$, and with this the necessary information for the first step of the Adams method has become available.

We shall explain this method of series expansion more specifically with the example of the Bessel equation. For this purpose, we begin with the Taylor series

$$y = a_0 + a_1 x + a_2 x^2 + a_3 x^3 + \ldots$$

with undetermined coefficients. Then $y(x_0) = y(0) = a_0$; i.e., we are looking for a solution which has a finite value at the singular initial point. Of course it is *a priori* questionable whether such a solution exists at all, because a singularity of the differential equation generally causes singularities of its solutions at the same point x_0. By differentiation it then follows that

$$y' = a_1 + 2a_2 x + 3a_3 x^2 + \ldots$$

$$y'' = 2a_2 + 6a_3 x + \ldots .$$

Substituting these three series in the differential equation (314) and comparing coefficients we obtain:

From the power x^{-1}: $a_1 = 0$

From the power x^0: $2a_2 = -2a_2 - a_0$, $a_2 = -\tfrac{1}{4}a_0$

From the power x^1: $6a_3 = -3a_3 - a_1$, $a_3 = -\tfrac{1}{9}a_1 = 0$.

The Taylor series therefore has the form

$$y = a_0(1 - \tfrac{1}{4}x^2 + \dots).$$

We choose $a_0 = 1$. This solution y is called the Bessel function of 0th order $J_0(x)$:

$$y = J_0(x) = 1 - \tfrac{1}{4}x^2 + \dots . \tag{315}$$

We shall now integrate the Bessel equation numerically by means of the second Runge-Kutta method. Since our solution y remains finite at the initial point $x_0 = 0$, it is not necessary to move the initial conditions to a neighboring point x_1 (as described above). Instead, we can proceed more simply as follows: The reduction to a system of first order reads:

$$y' = z, \qquad z' = -\frac{1}{x}z - y.$$

For the two functions f, g, needed in the computational scheme (310) we therefore obtain

$$f(x, y, z) = z, \qquad g(x, y, z) = -\frac{z}{x} - y. \tag{316}$$

Furthermore, we derive from the series (315) the initial conditions

$$y_0 = y(0) = 1, \qquad z_0 = y'(0) = 0. \tag{317}$$

For the start of the method we still need the two values $f(x_0, y_0, z_0)$ and $g(x_0, y_0, z_0)$. The first one presents no problems:

$$f(x_0, y_0, z_0) = z_0 = 0. \tag{318}$$

With the second one, the x in the denominator of the right-hand formula (316) causes some problems. For this reason we again make use of our series:

$$z = y' = -\tfrac{1}{2}x + \dots , \qquad z/x = -\tfrac{1}{2} + \dots .$$

For $x = 0$ we then have

$$z/x = -\tfrac{1}{2}, \qquad g(x_0, y_0, z_0) = \tfrac{1}{2} - 1 = -\tfrac{1}{2}. \tag{319}$$

Now all difficulties caused by the singularity have been removed; in Appendix I, Example 9, three integration steps have been executed using the scheme (310).

6.43   Linear Systems with Constant Coefficients

The differential equations (308) of our electrical oscillator are linear; moreover, they have constant coefficients since the capacitance C and the inductance L do not change during the oscillation. The general linear *homogeneous* system of differential equations with constant coefficients for two unknown functions $y(x)$, $z(x)$, has the form

$$y' = ay + bz, \qquad z' = cy + dz, \tag{320}$$

here a, b, c, d are given constants. Such systems can be solved in closed form by looking for solutions in the form of the exponential functions

$$y = ue^{-\lambda x}, \qquad z = ve^{-\lambda x}.$$

(We use the negative sign in the exponents since in the following we will be primarily interested in decaying solutions.) Substitution in (320) results in the equations

$$
\left.
\begin{aligned}
au + bv &= -\lambda u \\
cu + dv &= -\lambda v
\end{aligned}
\right| \tag{321}
$$

for the determination of u and v. This is a *special eigenvalue problem* in the sense of 5.2, and as such it belongs to the field of nonlinear algebra. This replacement of the original linear problem by a nonlinear one appears to be a questionable advantage if there are not just two, but many coupled differential equations to be solved. Our step by step methods therefore retain their significance—also in those cases which are explicitly integrable.

For our following discussion of the explicit solution, we make the assumption that the eigenvalue problem (321) has two different eigenvalues λ, λ', and that the corresponding nonproportional eigensolutions are u, v and u', v'. (The primes here have nothing to do with derivatives.) This results in two particular solutions

$$ue^{-\lambda x}, ve^{-\lambda x} \qquad \text{and} \qquad u'e^{-\lambda' x}, v'e^{-\lambda' x}$$

of the given system of differential equations. From these the general solution

$$y = \alpha u e^{-\lambda x} + \alpha' u' e^{-\lambda' x}, \qquad z = \alpha v e^{-\lambda x} + \alpha' v' e^{-\lambda' x} \tag{322}$$

is generated as a linear combination with constant coefficients α, α'. Let the initial conditions

$$y(0) = y_0, \qquad z(0) = z_0$$

be added to the differential equations (320). In order for the general solution to assume these values at $x = 0$, the equations

$$
\begin{aligned}
u\alpha + u'\alpha' &= y_0 \\
v\alpha + v'\alpha' &= z_0
\end{aligned}
\tag{323}
$$

must be satisfied. Solving them yields α, α'. It should be added here that in the symmetric case this process of solution can be simplified considerably ("symmetric" means that the matrix of the eigenvalue problem (321) is symmetric, i.e., that $b = c$). This can be seen by forming the Gauss normal equations corresponding to the equations (323) [3.1, footnote following formula (94)]. They read

$$
\begin{aligned}
(u^2 + v^2)\alpha + (uu' + vv')\alpha' &= uy_0 + vz_0 \\
(uu' + vv')\alpha + (u'^2 + v'^2)\alpha' &= u'y_0 + v'z_0
\end{aligned}
$$

Since λ, λ' are assumed to be different from each other, the two eigenvectors are orthogonal to each other, according to a theorem in 5.23; hence,

$$
uu' + vv' = 0
$$

and the normal equations yield

$$
\alpha = \frac{uy_0 + vz_0}{u^2 + v^2}, \qquad \alpha' = \frac{u'y_0 + v'z_0}{u'^2 + v'^2}.
\tag{323a}
$$

In each case the numerator contains the scalar product of the eigenvector with the vector y_0, z_0 of the initial conditions.

This result can be generalized easily to systems of n linear homogeneous differential equations with constant coefficients. In brief, we shall call the eigenvalues λ, λ', ... *the eigenvalues of the system of differential equations.*

6.44 Some Remarks about Numerical Stability

Let us now solve the given system of differential equations (320) by means of our numerical step by step methods. We choose a step h leading from $x_0 = 0$ to $x_1 = h$, and consider first only the rough Euler method. According to formula (293) in 6.31, this method yields the following approximate values y_1, z_1 for $y(x_1)$, $z(x_1)$:

$$
y_1 = y_0 + h(ay_0 + bz_0), \qquad z_1 = z_0 + h(cy_0 + dz_0).
$$

Adding further steps of the length h, we find the recursion formulas

$$
y_{n+1} = y_n + h(ay_n + bz_n), \qquad z_{n+1} = z_n + h(cy_n + dz_n)
\tag{324}
$$

for the successive approximate calculation of the function values at the points $x_n = nh$. However, we shall not be satisfied with this recursive method of calculation, but instead we derive explicit formulas for y_n, z_n, permitting a theoretical discussion of the course of the integration. Linear recursion formulas with constant coefficients can also be solved by exponential expressions. These are most conveniently assumed to have the form

$$y_n = u\rho^n, \qquad z_n = v\rho^n. \tag{325}$$

Now considering

$$y_{n+1} = u\rho^{n+1} = u\rho\rho^n$$

we find that upon substitution in (324) the power ρ^n drops out, leaving

$$u\rho = u + h(au + bv), \qquad v\rho = v + h(cu + dv)$$

or

$$\left. \begin{aligned} au + bv &= -\frac{1-\rho}{h}\, u \\[2mm] cu + dv &= -\frac{1-\rho}{h}\, v \end{aligned} \right\} \tag{326}$$

But this is exactly the eigenvalue problem (321), as can be seen by setting

$$\frac{1-\rho}{h} = \lambda, \qquad \rho = 1 - \lambda h. \tag{327}$$

The eigensolutions u, v of (326) are then identically equal to those of (321). For this reason we have denoted them with the same letters from the very beginning. The initial conditions are considered in the same way as in (322), and this provides us with the final result

$$\begin{aligned} y_n &= \alpha u(1 - \lambda h)^n + \alpha' u'(1 - \lambda' h)^n, \\ z_n &= \alpha v(1 - \lambda h)^n + \alpha' v'(1 - \lambda' h)^n \end{aligned} \tag{328}$$

where α, α' are the *same* coefficients as in (322). The formulas accordingly supply us with the results of the numerical integration without necessitating the actual execution of this integration. We can use this result in an obvious way to discuss the discretization error which will occur after *numerous* integration steps. This is more important (but also more difficult) than an estimation of the error committed in every single step. As an illustrative example, let us consider the differential equation

$$y'' + y = 0, \qquad y(0) = 0, \qquad y'(0) = 1.$$

Its exact solution is $y = \sin x$. Reduction to a system yields

$$y' = z, \qquad z' = -y, \qquad y_0 = 0, \qquad z_0 = 1.$$

We find that the eigenvalues are $\lambda = i$, $\lambda' = -i$ and the eigensolutions

$$u = 1, \quad v = -i \quad \text{and} \quad u' = 1, \quad v' = i.$$

The solution of (323) gives

$$\alpha = -\frac{1}{2i}, \qquad \alpha' = \frac{1}{2i}$$

and according to (328) the numerical integration produces

$$y_n = \frac{1}{2i} (1 + ih)^n - \frac{1}{2i} (1 - ih)^n. \tag{329}$$

In order to bring this into real form, we write the complex number $(1 + ih)$ in polar coordinates:

$$1 + ih = re^{i\varphi}, \qquad r = \sqrt{1 + h^2}, \qquad \tan \varphi = h.$$

For small h, therefore, r is only a little larger than 1. It now follows from (329) that

$$y_n = r^n \sin n\varphi.$$

Furthermore, φ is a little smaller than h; we therefore set $\varphi = h - \varepsilon$, and accordingly,

$$y_n = r^n \sin (nh - n\varepsilon) = r^{x_n/h} \sin \left(1 - \frac{\varepsilon}{h}\right) x_n.$$

Consequently, the numerical integration forms the curve

$$y = r^{x/h} \sin \left(1 - \frac{\varepsilon}{h}\right) x$$

instead of the curve $y = \sin x$. First of all, the numerically constructed curve oscillates higher; its maxima are becoming larger than 1 with the time. Secondly, it fails to keep time since it oscillates with the wrong frequency $(1 - \varepsilon/h)$.

We have entered into these considerations more as a side issue, and we shall now turn to the actual purpose of this section. For reasons of demonstration we shall use the differential equation

$$y'' - y = 0, \qquad y(-1) = 1, \qquad y'(-1) = -1.$$

This means that we use as initial point $x_0 = -1$. Obviously, the exact solution is

$$y = e^{-x-1} \tag{330}$$

and for large, positive x this function tends to zero. By proceeding with the numerical integration toward the right, we arrive after a few steps at the point $x = 0$. Here y and y' have the exact values

$$y(0) = \frac{1}{e}, \qquad y'(0) = -\frac{1}{e} \qquad \left(\frac{1}{e} = 0.36788 \dots\right).$$

Inasmuch as any desk-calculator as well as any automatic digital computer can only calculate with a finite number of places, not even the finest integration methods can give us these values. Instead, we obtain results which are subject to small round-off errors. For example, let us assume that a calculation with four-place accuracy results in

$$y(0) = 0.3679, \qquad y'(0) = -0.3677. \tag{331}$$

The subsequent course of our integration procedure now consists of numerically integrating the differential equation, starting at the point $x_0 = 0$ with the initial conditions (331). We shall use our formula (328) to determine what this will produce. Calculations analogous to those in the previous example yield the eigenvalues

$$\lambda = 1, \qquad \lambda' = -1$$

and the corresponding eigenvectors $u = 1$, $v = -1$, and $u' = 1$, $v' = 1$, respectively. The equations (323) have the form

$$\alpha + \alpha' = 0.3679$$
$$-\alpha + \alpha' = -0.3677$$

and give

$$\alpha = 0.3678, \qquad \alpha' = 0.0001.$$

As a result, (328) becomes

$$y_n = 0.3678(1 - h)^n + 0.0001(1 + h)^n. \tag{332}$$

Because of the second term, this expression will become infinitely large for increasing n, while, on the other hand, the solution of our differential equation decreases. Because the coefficient 0.0001 is small, this perturbation will naturally show up rather late, but after sufficiently many steps the integration will be thoroughly spoiled just the same. This phenomenon is called *numerical instability*. Note that it is not the result of an accumulation of round-off errors. In fact, the formula (332) describes the integration process under the assumption that the calculation is being performed exactly. Instead, only the small inaccuracy in the initial condition (331) is responsible for the instability. As soon as an inaccuracy due to round-off occurs but once during the integration process, this instability will result.

Incidentally, we will also encounter instability in the previous example if starting from $x = 0$ the integration is performed with a finer method which has a smaller discretization error for every single step than does the Euler method. Even if no discretization error occurs at all, the instability will remain, due to the fact that, according to (332), the exact solution of the differential equation under the initial condition (331) has the form

$$y = 0.3678e^{-x} + 0.0001e^x. \tag{333}$$

Here the first term is the sought function (330), (up to small deviations in the fourth decimal); the second term causes the instability.

In our example, the instability is therefore not caused by the numerical integration method used; it is *inherent* in the given problem. However, there are also cases where a suitably chosen numerical method remains stable for the same problem while another method introduces instabilities. This then is the actual *numerical* instability for which we will give an example immediately below. Unfortunately, it so happens that exactly those methods tend to be unstable which have a small discretization error for the individual step.

The result (333) also brings out the deeper reason for the inherent instability. It will always occur at the time when the decreasing solution must be constructed for a differential equation which also possesses strongly increasing solutions. Fortunately, the following technique exists for obtaining a stable computational method in spite of this. We first solve the differential equation *asymptotically*—i.e., through analytical means we derive formulas which permit the computation of the desired decreasing solution $y(x)$ and of its derivative for large x with a high degree of accuracy. Then we choose a large abscissa x_0 as initial point and from the asymptotic formulas we calculate the initial values $y(x_0)$, $y'(x_0)$. Starting with x_0, we now integrate numerically *toward the left* (i.e., toward decreasing x). This process has a good chance of remaining stable. Asymptotic methods are an important tool in numerical mathematics for many other problems as well.

Finally, we shall prepare a result here to be used later on. Let both eigenvalues λ, λ' of the system of differential equations (320) be real and *positive* numbers. Then the solution (322) will decrease for increasing x, no matter what the initial conditions are. Under these assumptions our Euler integration method will make sense only if the results y_n, z_n of the formula (328) also tend to zero for increasing n, and this for any arbitrary choice of α, α'. This results in the conditions

$$|1 - \lambda h| < 1, \qquad |1 - \lambda' h| < 1.$$

Since λ, λ' are positive, this is equivalent to

$$1 - \lambda h > -1, \qquad 1 - \lambda' h > -1$$

or, in short

$$h < \frac{2}{\lambda_{\max}} \tag{334}$$

where $\lambda_{\max}$ denotes the largest eigenvalue. This stability condition for the step length remains exactly the same when we are concerned with a system of n linear differential equations (with constant coefficients) whose eigenvalues are all positive.

We shall derive a similar stability condition when, instead of the Euler method, the integration is performed by means of the trapezoidal rule which we used to solve our introductory example of the electric oscillator circuit. In place of the formula (324) we then have the rules

$$y_{n+1} = y_n + \frac{h}{2}[(ay_n + bz_n) + (ay_{n+1} + bz_{n+1})]$$

$$z_{n+1} = z_n + \frac{h}{2}[(cy_n + dz_n) + (cy_{n+1} + dz_{n+1})].$$

This is a linear system of equations for y_{n+1}, z_{n+1}. Rather than solving it, we introduce (325) directly and in this way arrive at the eigenvalue problem

$$au + bv = -\frac{2}{h}\frac{1-\rho}{1+\rho}u$$

$$cu + dv = -\frac{2}{h}\frac{1-\rho}{1+\rho}v.$$

Comparison with (321) yields

$$\frac{2}{h}\frac{1-\rho}{1+\rho} = \lambda, \qquad \rho = \frac{2-h\lambda}{2+h\lambda}.$$

Because of (325) the stability condition therefore reads $\rho < 1$ or

$$\left|\frac{2-h\lambda}{2+h\lambda}\right| < 1.$$

But for positive λ this condition is always satisfied. Hence, we have discovered the following theorem: For a linear system of differential equations with constant coefficients and positive eigenvalues the integration by means of the *trapezoidal rule is always stable.* With this result we have already found an example for the actual numerical instability. When applied to the same problem, one method (trapezoidal rule) is always stable while the other (the Euler method) can cause instability by violating condition (334).

In general, most methods which involve the solution of a system of equations for each integration step are more stable than those methods which give the function values in an explicit form at the end of the step. A theory of numerical stability can be found in [35].

6.45 The Method of Steepest Descent

In Section 4.41 we introduced the concept of the definite nonlinear problem and we also described geometrically the solution of the problem by the method of steepest descent (Fig. 18). Continuing with this, we shall now work out the computational rules for such a gradient method.

Consider a function $F(x, y)$ for which the minimum is to be determined. After choosing an arbitrary initial point $P_0(x_0, y_0)$ in the x, y plane, we have to construct that orthogonal trajectory c to the level lines of $F(x, y)$ which originates from P_0 (Fig. 18). At an arbitrary point $P(x, y)$ the gradient of F has the components

$$F_x = \frac{\partial F}{\partial x}, \qquad F_y = \frac{\partial F}{\partial y}. \tag{335}$$

This vector is orthogonal to the level line through P and points in the direction of increasing values of F. Hence, the tangent vector to the curve c must have the components

$$-pF_x, \qquad -pF_y.$$

Here p is a proportionality factor which is essentially positive and which may vary along the trajectory. The differential equations of c consequently read

$$\frac{dx}{dt} = -p(t)F_x(x, y), \qquad \frac{dy}{dt} = -p(t)F_y(x, y). \tag{336}$$

where t is a parameter along the curve c; the function $p(t)$ must be continuous and positive, but can otherwise be chosen arbitrarily. The different gradient methods are distinguished by the choice of this function and by the numerical method used for the integration of the system of differential equations (336). Here a coarse numerical method is usually preferable, because there is no interest in determining the exact course of the orthogonal trajectory. Instead we are interested only in constructing, via a step by step method, a sequence

P_n of points which reproduce in some way the course of c, and which further-more has the essential property that $F(x, y)$ decreases when we proceed from P_n to P_{n+1}. We therefore choose the simple Euler integration method (293). In the case of our system (336) it provides us with the rule

$$x_1 = x_0 - hp(0)F_x(x_0, y_0), \qquad y_1 = y_0 - hp(0)F_y(x_0, y_0).$$

If a constant step length h is used for the continuation of the integration, the following rule for the step from P_n to P_{n+1} results:

$$x_{n+1} = x_n - hp(nh)F_x(x_n, y_n), \qquad y_{n+1} = y_n - hp(nh)F_y(x_n, y_n). \quad (337)$$

Geometrically, this simply means that we proceed from P_n along the tangent for a little while and that we therefore construct a *polygon*[1] instead of the curve c. Instead of choosing h and $p(t)$, we can also choose the (positive) numbers

$$\frac{1}{q_n} = hp(nh)$$

directly, and then use the computational rule

$$x_{n+1} = x_n - \frac{1}{q_n} F_x(x_n, y_n), \qquad y_{n+1} = y_n - \frac{1}{q_n} F_y(x_n, y_n). \quad (338)$$

Any iterative procedure for the solution of the given minimization problem is also called a *relaxation method*; (338) is the relaxation formula and the numbers q_n are called relaxation coefficients. The method's success depends on their appropriate choice.

Let us first discuss the special choice $p(t) \equiv 1$. The relaxation formula (337) then has the form:

$$x_{n+1} = x_n - hF_x(x_n, y_n), \qquad y_{n+1} = y_n - hF_y(x_n, y_n). \quad (339)$$

This is the Euler method for the solution of the differential equations

$$\frac{dx}{dt} = -F_x, \qquad \frac{dy}{dt} = -F_y \quad (340)$$

of the orthogonal trajectory c. In order to be more explicit, we shall further assume that $F(x, y)$ is a quadratic function

$$2F(x, y) = Ax^2 + 2Bxy + Cy^2$$
$$+ 2kx + 2ly \quad (341)$$

---

[1] For clarity's sake the Euler polygon in Fig. 18 has been drawn as originating from a different initial point.

with a *positive definite* quadratic form in the first row. Figure 18 then becomes the system of concentric ellipses of Fig. 8. We have

$$F_x = Ax + By + k, \qquad F_y = Bx + Cy + l. \tag{342}$$

The differential equations (340) read

$$\frac{dx}{dt} = -(Ax + By + k), \qquad \frac{dy}{dt} = -(Bx + Cy + l) \tag{343}$$

and the relaxation formula (339) assumes the form

$$x_{n+1} = x_n - h(Ax_n + By_n + k), \qquad y_{n+1} = y_n - h(Bx_n + Cy_n + l). \tag{344}$$

The minimum of F is assumed at the point (x, y) where $F_x = 0$, $F_y = 0$. This point is also the solution of the *symmetric* system of equations

$$\left. \begin{aligned} Ax + By + k = 0 \\ Bx + Cy + l = 0 \end{aligned} \right| \tag{345}$$

As will be shown below, the sequences x_n, y_n tend respectively to the solution x, y of this system of equations, provided that h is chosen sufficiently small. Furthermore, the values in the brackets of (344) are the *residuals* which are obtained when the nth point $P_n(x_n, y_n)$ of the relaxation sequence is substituted in the equations (345).

With this we have found a method of solving by means of relaxation a given linear and symmetric system of equations (345) (resulting from a definite problem). This is called the *Richardson* method. S. Frankel, as well as D. A. Flanders and G. Shortly, have improved it to obtain a faster convergence of the relaxation sequence x_n, y_n [16].

For a system of linear equations

$$\sum_{k=1}^{n} a_{ik}x_k + c_i = 0, \qquad i = 1, 2, \dots, n$$

with more than two unknowns we can proceed in a similar manner. Let the system be symmetric and definite. This means first of all that $a_{ik} = a_{ki}$ and secondly that the system is equivalent to the minimization of the quadratic function

$$2F = \sum_{(i,k)} a_{ik}x_ix_k + 2\sum_{(i)} c_ix_i$$

where the first term quadratic form is positive definite. Instead of (343) we then have the system of differential equations

$$\frac{dx_i}{dt} = -\left(\sum_{k=1}^{n} a_{ik}x_k + c_i \right). \tag{346}$$

Denoting the νth approximation of x_i by $x_i^{(\nu)}$, we obtain the relaxation formula

$$x_i^{(\nu+1)} = x_i^{(\nu)} - h\left(\sum_{k=1}^{n} a_{ik}x_k^{(\nu)} + c_i\right),\qquad(347)$$

where h is a chosen step length.

However, it is advisable to apply these and other relaxation methods for the solution of *linear* equations only when the matrix of the given system of equations contains many zeros. This is the case especially when the given equations are the result of discretization of a boundary value problem (6.52). In that case, relaxation methods are of the greatest usefulness, particularly when automatic computers are being used. For all other cases the Gauss algorithm is still the best method of solving linear equations. Of course, in the *nonlinear* case where F is not a quadratic function, no Gauss algorithm is available and relaxation is often the last and only means of solving a given problem.

With respect to the choice of the step length h in the method (344), (347) of *Richardson*, the following considerations can be added. We first consider the homogeneous case $k = 0$, $l = 0$. The solution of the given equations (345) is $x = 0$, $y = 0$; furthermore, the system of differential equations (343) has the form

$$\frac{dx}{dt} = -Ax - By, \qquad \frac{dy}{dt} = -Bx - Cy.$$

It is linear and has constant coefficients and can therefore be solved by means of the formulas (320) through (323) of the previous section. Using the notation (320) we have

$$a = -A, \qquad b = -B, \qquad c = -B, \qquad d = -C.$$

Hence, the equations (321) for the determination of the eigenvalues of the system of differential equations read:

$$\left.\begin{array}{l} -Au - Bv = -\lambda u \\ -Bu - Cv = -\lambda v \end{array}\right| \quad \text{or} \quad \left.\begin{array}{l} Au + Bv = \lambda u \\ Bu + Cv = \lambda v \end{array}\right| \qquad (348)$$

The matrix on the right is simply the matrix of the given equations (345). We shall therefore call the λ in brief *the eigenvalues of the given system of equations* (345). According to (322), the solutions of the differential equations then are:

$$x = \alpha u e^{-\lambda t} + \alpha' u' e^{-\lambda' t}, \qquad y = \alpha v e^{-\lambda t} + \alpha' v' e^{-\lambda' t},$$

where the constants α, α' depend on the choice of the initial point P_0. Since the eigenvalue problem (348) is symmetric, the eigenvalues λ, λ' are real (5.23). Moreover, λ, λ' are even > 0, because for increasing t the orthogonal trajectory must tend to the point of solution $x = 0$, $y = 0$, (for any choice of the initial point P_0).[1] Hence we find ourselves in precisely the same situation we discussed at the conclusion of the previous section on numerical stability. In order that the Euler method (i.e., the method of *Richardson*) decreases toward its solution $x = 0$, $y = 0$, the condition (334) must be satisfied:

$$h < \frac{2}{\lambda_{\max}}. \tag{349}$$

Here $\lambda_{\max}$ is the largest eigenvalue of the given linear system of equations. In the general case when k, l are not equal to zero, exactly the same result will hold. In order to see this we need only subject the x, y-coordinate system to a parallel translation in such a way that the origin coincides with the point represented by the solution of (345).

To satisfy (349), a coarse but safe upper bound of $\lambda_{\max}$ will suffice. For this purpose the *Gerschgorin* theorem can well be used (5.24). It is advisable not to choose the step length too small since that would prevent one from getting ahead properly on the trajectory.

Up to now we have assumed the function $p(t)$ in (336) to be identically $= 1$. The following choice also has its advantages:

$$p(t) = \frac{1}{F_x^2 + F_y^2}.$$

The differential equations (336) then have the form

$$\frac{dx}{dt} = -\frac{F_x}{F_x^2 + F_y^2}, \qquad \frac{dy}{dt} = -\frac{F_y}{F_x^2 + F_y^2}$$

and the Euler integration method leads to the recursion formulas

$$x_{n+1} = x_n - h\frac{F_x}{F_x^2 + F_y^2}, \qquad y_{n+1} = y_n - h\frac{F_y}{F_x^2 + F_y^2}. \tag{350}$$

Here the partial derivatives F_x, F_y have to be taken at the point x_n, y_n.

---

[1] For linear equations with more than two unknowns we can of course not call upon geometric intuition to assert that a curve of steepest descent leading into a bowl-shaped valley does reach the deepest point of this valley. In this case, we must reason as follows: The λ are also the eigenvalues of the quadratic form constituting the first part of the quadratic function F. Since this form was assumed to be positive definite the λ will be positive.

For the variation of the function F along the trajectory it follows that

$$\frac{dF}{dt} = F_x \frac{dx}{dt} + F_y \frac{dy}{dt} = -F_x \frac{F_x}{F_x^2 + F_y^2} - F_y \frac{F_y}{F_x^2 + F_y^2} = -1$$

and by integration that

$$F(x, y) = F_0 - t,$$

where F_0 is the value of F at the initial point P_0. The parameter t is now no longer an auxiliary variable without deeper significance for the problem, but it is essentially the function F itself. After one relaxation step (350) it can be expected that $F(x, y)$ has decreased by approximately h, provided that h has been chosen sufficiently small. The method is especially convenient in those occasional cases where the minimal value $F_{\min}$ of $F(x, y)$ is known beforehand (or can at least be estimated), but where the point (x, y) is still to be found in which this value is assumed. If one has enough time and money to calculate N relaxation steps on an automatic computer, one will choose

$$h = \frac{F_0 - F_{\min}}{N},$$

or one first chooses h much larger, calculates until the relaxation stops (i.e., until F increases in the final step), and continues then with a considerably smaller h. For this method (350) the relaxation coefficients are equal to

$$q_n = \frac{1}{h} [F_x(x_n, y_n)^2 + \bar{F}_y(x_n, y_n)^2],$$

i.e., they change from step to step. Hence the computational rules (350) belong to the general circle of methods where the relaxation coefficients are derived from the information available at the nth step about the function F and its derivatives. In the linear case (F = quadratic function) again finer methods can be developed, including the method of conjugate gradients of *M. Hestenes* and *E. Stiefel* [16].

Example. Let us assume that the two sides of a rectangle have been measured and the values 3 and 4 obtained for them. Furthermore, the measurement of a diagonal gave 6. The problem is to find for these sides values x, y which best fit the three measured data. The theory of least squares approximation yields the following error equations:

$$\left. \begin{array}{r} x \qquad - 3 = r_1 \\[2mm] y \; - 4 = r_2 \\[2mm] \sqrt{x^2 + y^2} - 6 = r_3. \end{array} \right|$$

On the right we have the residuals. The problem could now be attacked by linearizing the error equations and forming the corresponding Gauss normal equations. However, this time we shall proceed directly according to the method of least-squares by requiring that

$$F(x, y) = r_1^2 + r_2^2 + r_3^2 = (x - 3)^2 + (y - 4)^2 + [\sqrt{x^2 + y^2} - 6]^2$$

is to become minimal. Then

$$F_x = 4x - 6 - \frac{12x}{\sqrt{x^2 + y^2}}, \qquad F_y = 4y - 8 - \frac{12y}{\sqrt{x^2 + y^2}}.$$

We start with the not particularly suitable initial point $x = 4$, $y = 4$. Then

$$F = 1.1177, \qquad F_x = 1.5147, \qquad F_y = -0.4853, \qquad F_x^2 + F_y^2 = 2.5298.$$

Here we now have one case where $F_{\min}$ can be estimated: F is strictly positive and can at most decrease to zero. If at most 10 relaxation steps are to be used to bring F from the initial value $F_0 \sim 1$ into the neighborhood of zero, about $h = 0.1$ should be chosen. The formulas (350) yield

$$x_1 = 4 - 0.1 \frac{1.5147}{2.5298} = 3.9401, \qquad y_1 = 4 + 0.1 \frac{0.4853}{2.5298} = 4.0192.$$

The corresponding value of F is

$$F_1 = 1.0223,$$

which shows that F has indeed decreased by about $h = 0.1$. Six further steps with constantly decreasing F bring us to

$$x_7 = 3.359, \qquad y_7 = 4.354, \qquad F_7 = 0.505,$$

while the next step yields $F_8 \sim 0.9$, thereby overshooting the minimum by far. Incidentally, the exact solution of the problem is

$$x = 3.3, \qquad y = 4.4, \qquad F = 0.5.$$

It does not really matter much if a numerical mistake is made at any point during this lengthy calculation—i.e., if a wrong point $P_n(x_n, y_n)$ is computed. During the further course of the calculation a new trajectory is obtained, originating from P_n and serving the same purpose as the old one. In view of this we say that gradient methods are *self-correcting*.

6.5 Boundary-Value Problems

In ballistics and astronautical problems it is not only essential to follow computationally the course of a projectile or a spacecraft sent out from a given point, but it is also essential to hit a given target at the end of the trajectory. Consequently, not only the initial conditions—as, up to now, in 6.3, 6.4—are given for the differential equations of the trajectory, but also the conditions at the end of the interval of the independent variables. And in that case we speak of a boundary-value problem. In our treatment of the linear oscillator in 5.41 we already met such boundary-value problems as a means for calculating oscillation frequencies.

6.51   The Method of Particular Solutions

As we did with the initial-value problems of 6.3, 6.4, we will start here with *linear* differential equations. For instance, consider the differential equations of second order

$$y'' = p(x)y + q(x)y' + r(x). \tag{351}$$

$y(x)$ is the unknown function, $p(x)$, $q(x)$, $r(x)$ are given functions of the independent variable x. The latter is to range from 0 to a and we assume that the following boundary conditions have been given:

$$y(0) = 0, \qquad y(a) = b. \tag{352}$$

If we make a drawing of this situation in an x, y-coordinate system, then a curve will have to be drawn from the origin. This curve will, first of all, have to satisfy the differential equation and secondly it must hit the target —namely, the point with the coordinates (a, b). In order to solve this problem with the techniques of 6.4, we first construct by numerical step by step integration a solution $Y(x)$ of the differential equation, satisfying the initial conditions

$$Y(0) = 0, \qquad Y'(0) = 1. \tag{353}$$

At first we satisfy only the left-hand boundary condition (352), but, in order to be able to start a numerical integration at all, we need an initial condition for the derivative, and this condition was chosen arbitrarily in (353).

We likewise compute by numerical integration a solution $\eta(x)$ of the homogeneous differential equation

$$\eta'' = p(x)\eta + q(x)\eta' \tag{354}$$

under the initial conditions

$$\eta(0) = 0, \qquad \eta'(0) = 1.$$

The most general solution of (351) which satisfies the left-hand boundary condition $y(0) = 0$ then has the form

$$y(x) = Y(x) + \alpha\eta(x).$$

Here α is a coefficient which can be chosen arbitrarily. We must only make certain that this curve hits the target. This condition leads to the linear equation

$$Y(a) + \alpha\eta(a) = b$$

for the determination of α.

Of course, more complicated boundary conditions can be prescribed as well. We consider the quite general case where the boundary conditions have the form

$$c_{11}y(0) + c_{12}y(a) + c'_{11}y'(0) + c'_{12}y'(a) = l_1$$
$$c_{21}y(0) + c_{22}y(a) + c'_{21}y'(0) + c'_{22}y'(a) = l_2. \tag{355}$$

Here the c and l are given numbers. Again, we construct a particular solution $Y(x)$ of (351). In addition, we now calculate two solutions $\eta_1(x)$, $\eta_2(x)$ of the homogeneous equation (354), satisfying, for example, the initial conditions

$$\eta_1(0) = 0, \qquad \eta'_1(0) = 1$$
$$\eta_2(0) = 1, \qquad \eta'_2(0) = 0.$$

The general solution of (351) then is given by

$$y(x) = Y(x) + \alpha_1\eta_1(x) + \alpha_2\eta_2(x)$$

and substitution in the boundary conditions (355) yields a linear system of equations for α_1, α_2.

In this connection we note that the boundary conditions

$$y(0) = y(a), \qquad y'(0) = y'(a)$$

are a special case of the conditions (355); they require the construction of a *periodic* solution of the given differential equation (351).

We now come to the *nonlinear* case where, instead of (351), we have a general differential equation

$$y'' = f(x, y, y') \tag{356}$$

of second order. We restrict ourselves to the simple boundary conditions (352):

$$y(0) = 0, \qquad y(a) = b.$$

As in the section on algebra (Chapter 4), we shall try to solve the boundary-value problem by *linearization*. For this purpose, an approximate solution $Y(x)$ must already be known—i.e., a curve which almost hits the target

$$Y(0) = 0, \qquad Y(a) = b + \Delta b, \tag{357}$$

Here Δb is a small quantity. Such a curve can be determined experimentally if by numerical integration several curves $Y(x)$ are constructed which begin at the origin but which there have different slopes $Y'(0)$. It is thereby practical to use the "angle-halving" technique of the artillerist. This method begins with two curves $Y_1(x)$, $Y_2(x)$, one of which is above and the other below the target:

$$Y_1(a) > b, \qquad Y_2(a) < b.$$

A third curve with the initial slope

$$Y'(0) = \tfrac{1}{2}[Y_1'(0) + Y_2'(0)]$$

is then constructed, and so on.

Let us assume that by means of this or some other technique an approximate solution $Y(x)$ has been calculated and tabulated. Let its initial slope be $Y'(0) = \omega_0$. The linearization process is based on a study of the family of solutions of the given differential equation (356), which satisfy the initial conditions

$$y(0) = 0, \qquad y'(0) = \omega, \tag{358}$$

where ω is the parameter of the family and the problem is to find that value of ω for which the corresponding curve hits the target. The individual solutions of this family are functions $y(x, \omega)$ of x and ω. We have to investigate in what way the solution $y(x, \omega)$ varies when for fixed x the parameter ω is changed by a small quantity. In other words, we must calculate the derivative

$$\eta(x) = \frac{\partial y}{\partial \omega}. \tag{359}$$

This is done by partial differentiation of (356) with respect to ω:

$$\eta'' = \frac{\partial y''}{\partial \omega} = \frac{\partial f}{\partial \omega} = \frac{\partial f}{\partial y} \frac{\partial y}{\partial \omega} + \frac{\partial f}{\partial y'} \frac{\partial y'}{\partial \omega} = \frac{\partial f}{\partial y} \eta + \frac{\partial f}{\partial y'} \eta'.$$

In particular, we calculate the derivative (359) at the value ω_0 of the parameter corresponding to the known approximate solution $Y(x)$. In more complete form we then have

$$\eta'' = \frac{\partial f}{\partial y} (x, Y, Y')\eta + \frac{\partial f}{\partial y'} (x, Y, Y')\eta'. \tag{360}$$

Here the two partial derivatives are functions of x which can be evaluated since $Y(x)$ is available in table form. We therefore introduce the notations:

$$p(x) = \frac{\partial f}{\partial y} (x, Y, Y'), \qquad q(x) = \frac{\partial f}{\partial y'} (x, Y, Y'), \tag{361}$$

and (360) then becomes

$$\eta'' = p(x)\eta + q(x)\eta' \tag{362}$$

which is a *linear* differential equation for the determination of $\eta(x)$. It is called the *perturbation equation* or *first variation* corresponding to the given nonlinear differential equation (356). Differentiating the initial condition (358) we find that at the point $x = 0$

$$\frac{\partial y}{\partial \omega} = \eta(0) = 0, \qquad \frac{\partial y'}{\partial \omega} = \eta'(0) = 1. \tag{363}$$

These are the initial conditions which have to be considered with the perturbation equation. Hence, the perturbation equation can be solved and $\eta(x)$ becomes a known function.

If now ω changes from ω_0 to the neighboring value $\omega_0 + \Delta\omega$, then—because of (359)—the solution $y(x, \omega)$ will be subjected to the increment $\eta(x)\Delta\omega$. Consequently,

$$y(x) \sim Y(x) + \eta(x)\Delta\omega.$$

The given boundary condition $y(a) = b$ now yields

$$Y(a) + \eta(a)\Delta\omega \sim b$$

or, because of (357),

$$\Delta\omega = -\frac{1}{\eta(a)}\Delta b.$$

With this linearized correction for ω, a new solution of the given differential equation (356) is calculated and this curve will probably hit the target much better.

In order to practice constructing the first variation, we here add the following small example: Consider a satellite circling the earth; in its orbit-plane we introduce rectangular coordinates x, y (with the origin at the center of the earth). Under the assumption of a homogeneous and exactly spherical earth, the equations of motion for the satellite have the form

$$\frac{d^2x}{dt^2} = -c(x^2 + y^2)^{-\frac{3}{2}}x, \qquad \frac{d^2y}{dt^2} = -c(x^2 + y^2)^{-\frac{3}{2}}y. \tag{364}$$

Here c is a positive constant. A special solution is the circular orbit with the radius r

$$X = r\cos\sqrt{\frac{c}{r^3}}\,t, \qquad Y = r\sin\sqrt{\frac{c}{r^3}}\,t, \tag{365}$$

in which the satellite has the angular velocity

$$\omega_0 = \frac{1}{r}\sqrt{\frac{c}{r}}.$$

Hence, at the time $t = 0$ the velocity vector has the components 0, $r\omega_0$. Let us assume that due to inaccuracies during launching the initial velocity has the components 0, $r\omega$; i.e., that orbital velocity has been achieved with respect to its direction but not with respect to its magnitude. Here ω is again a variable parameter. If now, similar to the above considerations, we introduce the perturbation functions

$$\xi = \frac{\partial x}{\partial \omega}, \qquad \eta = \frac{\partial y}{\partial \omega},$$

then differentiation of the first equation (364) with respect to ω yields

$$\frac{d^2\xi}{dt^2} = [3c(x^2 + y^2)^{-\frac{5}{2}}x^2 - c(x^2 + y^2)^{-\frac{3}{2}}]\xi + [3c(x^2 + y^2)^{-\frac{5}{2}}xy]\eta.$$

For x, y we substitute here the circular trajectory (365), and after some simple transformations we obtain the linear perturbation equation of second order

$$\frac{d^2\xi}{dt^2} = \frac{\omega_0^2}{2}[(1 + 3\cos 2\omega_0 t)\xi + (3\sin 2\omega_0 t)\eta]. \tag{366}$$

A similar treatment of the second equation (364) leads to:

$$\frac{d^2\eta}{dt^2} = \frac{\omega_0^2}{2}[(3\sin 2\omega_0 t)\xi + (1 - 3\cos 2\omega_0 t)\eta]. \tag{367}$$

In this way, a system of two linear differential equations of second order has been obtained in which the coefficients are periodic functions of the independent variable t. There are extensive theories [31] about the integration of linear differential equations with *periodic coefficients*, in which Fourier series play an important role.

According to our definitions, the initial conditions for the system (364) have the form

$$x(0) = r, \qquad \frac{dx}{dt}\bigg|_{t=0} = 0, \qquad y(0) = 0, \qquad \frac{dy}{dt}\bigg|_{t=0} = r\omega.$$

Differentiation with respect to ω yields for the initial conditions of the perturbation equations

$$\xi(0) = 0, \qquad \frac{d\xi}{dt}\bigg|_{t=0} = 0, \qquad \eta(0) = 0, \qquad \frac{d\eta}{dt}\bigg|_{t=0} = r.$$

By chance, the solutions of the perturbation equations can be given in closed form here, while, generally, numerical integration must otherwise be used. We find

$$\xi(t) = r\left[3t\sin\omega_0 t + \frac{1}{\omega_0}(-3 + 2\cos\omega_0 t + \cos 2\omega_0 t)\right]$$

$$\eta(t) = r\left[-3t\cos\omega_0 t + \frac{1}{\omega_0}(2\sin\omega_0 t + \sin 2\omega_0 t)\right].$$

If the satellite has the perturbed angular velocity $\omega = \omega_0 + \Delta\omega$, then its orbit is given in linear approximation by

$$x = X(t) + \xi(t)\Delta\omega, \qquad y = Y(t) + \eta(t)\Delta\omega.$$

This problem could also have been handled by using the fact that the perturbed orbit is a Kepler ellipse.

6.52 Method of Discretization

We shall confine ourselves in the following to *linear* boundary-value problems. We show the fundamental principle of discretization with the example of the boundary value problem for the unknown function $y(x)$:

$$y'' = p(x), \qquad y(0) = 0, \qquad y(a) = b. \tag{368}$$

This problem can scarcely be surpassed in its simplicity; $p(x)$ is a given function. We divide the *fundamental interval* $(0, a)$ of the boundary value problem into n equal parts of length h by introducing the partition points:

$$x_k = kh, \qquad h = \frac{a}{n}, \qquad k = 0, 1, \dots, n.$$

Next we introduce the values

$$y_k = y(x_k), \qquad p_k = p(x_k)$$

of the unknown and of the given function at these partition points, contenting ourselves with calculating only the interpolating values y_k (see 6.1) rather than the entire course of the function $y(x)$. Furthermore, we approximate all derivatives appearing in the differential equation (or possibly in the boundary conditions) by values which are the result of numerical differentiation[1] based on the I-values (*difference method*). Using the rule for numerical differentiation (6.1) or the formula (263) we obtain in our case

$$y''(x_k) \sim \frac{y_{k-1} - 2y_k + y_{k+1}}{h^2}, \qquad k = 1, 2, \dots, n - 1.$$

Consequently, the boundary-value problem can be replaced approximatively by the system of linear equations

$$y_{k-1} - 2y_k + y_{k+1} = h^2 p_k, \qquad k = 1, 2, \dots, (n-1) \tag{369}$$

[1] We must be on guard against the temptation to use such a technique of numerical differentiation with the initial-value problems of 6.3 and 6.4; catastrophic numerical instabilities could at times result. In the case of boundary-value problems we need not be afraid of such instabilities since the unknown function is "nailed down" at both ends of the fundamental interval by the boundary conditions.

for the unknown $y_1, y_2, \ldots, y_{n-1}$ (y_0 and y_n being given by the boundary conditions). The finer the partition has been made, the better the approximation will be. For $n = 8$ this system of equations reads in full:

$$
\begin{aligned}
-2y_1 + \ y_2 &= h^2 p_1 \\
y_1 - 2y_2 + \ y_3 &= h^2 p_2 \\
y_2 - 2y_3 + \ y_4 &= h^2 p_3 \\
y_3 - 2y_4 + \ y_5 &= h^2 p_4 \\
y_4 - 2y_5 + \ y_6 &= h^2 p_5 \\
y_5 - 2y_6 + \ y_7 &= h^2 p_6 \\
y_6 - 2y_7 &= h^2 p_7 - b
\end{aligned}
\tag{370}
$$

The boundary conditions have been used in the first and last equation. We have taken up all this space to bring out the special structure of the system of equations. It has very many unknowns (n can be large), but every equation contains at most three unknowns. The matrix of the system of equations has very many zeros; the elements different from zero can be arranged in a band leading from the left-hand upper corner to the right-hand lower one. Such *band matrices* always occur when a boundary-value problem is subjected to discretization, also when the given differential equation is a partial differential equation.

The equations (369) can be solved by means of the Gauss algorithm. This computational method has the very welcome property of not destroying the band structure during the elimination process, provided that the pivots in each case are chosen in the left-hand upper corner of the Gauss table. This follows directly from the rectangle rule (1.1). But we must still prove here that the mentioned choice of pivots is permitted in the example under discussion. For this purpose, we introduce the following quadratic function of the variables $y_1, y_2, \ldots, y_{n-1}$:

$$
F = y_1^2 + (y_1 - y_2)^2 + (y_2 - y_3)^2 + \ldots + (y_{n-2} - y_{n-1})^2 + (y_{n-1} - b)^2
$$
$$
+ 2h^2 p_1 y_1 + 2h^2 p_2 y_2 + \ldots + 2h^2 p_{n-1} y_{n-1}.
$$

This function is to be minimized. Differentiating with respect to y_k and considering the boundary conditions $y_0 = 0$, $y_n = b$, we obtain the linear equations:

$$
-y_{k-1} + 2y_k - y_{k+1} + h^2 p_k = 0
\tag{371}
$$

which are equivalent to the equations (369). The function F begins with the quadratic form

$$y_1^2 + (y_1 - y_2)^2 + \ldots + (y_{n-2} - y_{n-1})^2 + y_{n-1}^2.$$

Since this form is positive definite, the system of equations (371) is symmetric definite. According to a result in 3.3, the indicated choice of the pivots can therefore never lead to a zero pivot.

Incidentally, the difference method functions in exactly the same way if, instead of a boundary-value problem, we have to solve an *eigenvalue problem*, as for example:

$$y'' = \lambda y, \qquad y(0) = 0, \qquad y(a) = 0.$$

Here we have to find a function $y(x)$ which is *not identically equal to zero* and which satisfies these conditions. In order to perform the discretization we need only replace $p(x)$ by $\lambda y(x)$. Instead of (369), we then obtain an eigenvalue problem

$$y_{k-1} - 2y_k + y_{k+1} = \lambda h^2 y_k, \qquad k = 1, 2, \ldots, (n-1),$$

as considered in our Chapter 5 ($y_0 = y_n = 0$).

Actually, we have already carried out these considerations a number of times, namely, when discussing the different static and dynamic forms of the problem of mass-points connected by elastic springs. For example, we refer to the equations (116) and (207). All this material merely appears here in a slightly different light.

Without doubt, the difference method is the most direct method for transforming a continuous problem into an algebraic one through discretization. As a result of its suitability for use on automatic computers, this method is becoming increasingly important. As a rule, the desk calculator is not capable of handling the enormous number of unknowns in the linear equations.

Only the coarsest rule has been used here for the numerical differentiation. Finer approximations have been developed, especially in [21]. The user of automatic computers will usually choose a very small step h rather than calculate with very fine approximation formulas. In the case of partial differential equations, such more sophisticated formulas are often undesirably complicated (especially toward the boundary of the fundamental domain).

We now turn to the solution of linear equations of the band type by means of *relaxation techniques*. For this purpose, we make the following assumptions about our boundary-value problem (368). Let

$$p(x) = \text{const} = 0.4, \qquad a = 5, \qquad b = 5,$$

so that the boundary-value problem now has the form

$$y'' = 0.4, \qquad y(0) = 0, \qquad y(5) = 5. \tag{372}$$

Obviously, the exact solution is the quadratic function

$$y = \tfrac{1}{5}x^2; \tag{373}$$

$h = 1$ shall be chosen for the discretization with the result that we have four inside partition points in the fundamental interval and correspondingly four linear equations (369)

$$y_{k-1} - 2y_k + y_{k+1} = 0.4, \qquad k = 1, 2, 3, 4, \tag{374}$$

whereby $y_0 = 0$, $y_5 = 5$, because of the boundary conditions.   Here the kth equation is obtained through numerical differentiation at the kth partition point.   The unknown y_k, corresponding to this partition point, is called the *central unknown* of this kth equation.   It is characterized by the fact that it appears on the diagonal of the system of equations, as illustrated by the explicitly written example (370).

As first relaxation method we discuss the so-called *single-step* procedure, also called the *Gauss-Seidel* method.   It makes use of the relation

$$y_k = \frac{y_{k-1} + y_{k+1}}{2} - 0.2, \tag{375}$$

obtained by solving the kth equation with respect to the central unknown. This relation states that for the final solution of the boundary-value problem the value y_k has to be 0.2 less than the arithmetic mean of its neighbors. We start out with reasonable approximation values for the y_k and successively improve them by forming this mean.   The details of this procedure are best explained with the help of the following table:

$x =$	0	1	2	3	4	5	
$y =$	0	1	2	3	4	5	(376)
	0	0.8	1.7	2.65	3.625	5	
	0	0.65	1.45	2.3375	3.46875	5	

The second row contains the initial approximation; we simply chose $y_k = x_k$. The third row has been calculated from left to right; the boundary value $y_0 = 0$ stands at the left end.   The next value is obtained through the above-mentioned evaluation of the mean:

$$y_1 = \frac{0 + 2}{2} - 0.2 = 0.8.$$

It is characteristic for this method that the next mean value already uses this improved value of y_1:

$$y_2 = \frac{0.8 + 3}{2} - 0.2 = 1.7.$$

In this way, the third row is calculated, after which the fourth row is started at the left end, and so on. This shows how the initial approximation is slowly "bent" into the parabola (373).

At this time, we also want to give the computational rules for applying the single-step method to an arbitrary system of equations:

$$\sum_{k=1}^{n} a_{ik}x_k + c_i = 0, \qquad i = 1, 2, \dots, n. \tag{377}$$

It makes no difference whether this system is derived from a boundary-value problem or not. The solution of the ith equation with respect to the central unknown x_i yields:

$$x_i = -\frac{1}{a_{ii}}\left[c_i + \sum_{k \neq i} a_{ik}x_k\right]$$

or

$$x_i = -\frac{1}{a_{ii}}\left[c_i + \sum_{k=1}^{i-1} a_{ik}x_k + \sum_{k=i+1}^{n} a_{ik}x_k\right].$$

If the various approximations are denoted by an upper index ν, the computational rule has the form

$$x_i^{(\nu+1)} = -\frac{1}{a_{ii}}\left[c_i + \sum_{k=i}^{i-1} a_{ik}x_k^{(\nu+1)} + \sum_{k=i+1}^{n} a_{ik}x_k^{(\nu)}\right]. \tag{378}$$

For each ν we begin with $i = 1$ and proceed until $i = n$. In our example (376), the index ν numbers the rows.

The special case where the given system of equations (377) is symmetric definite is of particular importance. It can be proven that in this special case the single-step method *converges* to the solution x_k of the system of equations. This does not need to be the case for arbitrary systems (377).

It should be stressed once again that, as an iterative method, the single-step method too can compete with the Gauss algorithm only when the given equations (377) are markedly of the band type. And even then the method pays off only when automatic computers are used. In general, the convergence is slow, as example (376) already illustrates. For symmetric-definite systems of equations, obtained through discretization of boundary-value problems with the help of the difference method, D. Young [16] has developed the so-called methods of *over-relaxation*. These methods improve the convergence considerably, and they are frequently used today for the solution of boundary-value problems on automatic computers.

In order to describe briefly the method of over-relaxation, we write (378) in the form:

$$x_i^{(\nu+1)} = x_i^{(\nu)} - \frac{1}{a_{ii}}\left[c_i + \sum_{k=1}^{i-1} a_{ik}x_k^{(\nu+1)} + \sum_{k=i}^{n} a_{ik}x_k^{(\nu)}\right]. \qquad (378a)$$

Here we have included the term for $k = i$ in the second sum and compensated for this by addition of $x_i^{(\nu)}$. The new approximation $x_i^{(\nu+1)}$ is then obtained from the old one, $x_i^{(\nu)}$, by addition of a correction term. This correction term contains in the square brackets the residual of the ith equation (377) obtained by substituting for the unknown those approximation values which are the best ones at this moment (compare 1.33).

The method of over-relaxation now consists in multiplying this correction term by a fixed universal *over-relaxation factor* $\omega > 1$, i.e., by over-stressing the correction. In other words, we calculate according to the equation

$$x_i^{(\nu+1)} = x_i^{(\nu)} - \frac{\omega}{a_{ii}}\left[c_i + \sum_{k=1}^{i-1} a_{ik}x_k^{(\nu+1)} + \sum_{k=i}^{n} a_{ik}x_k^{(\nu)}\right]. \qquad (378b)$$

It is difficult to give general rules for the choice of ω. In the case of a symmetric-definite system (377), the condition $\omega < 2$ has to be observed if the method is to remain convergent.

For purposes of illustration, we shall solve our equations (374) by over-relaxation with $\omega = 1.4$. While table (376) was calculated by use of the formula

$$y_k^{(\nu+1)} = \frac{y_{k-1}^{(\nu+1)} + y_{k+1}^{(\nu)}}{2} - 0.2 \qquad (375a)$$

$$= y_k^{(\nu)} + \tfrac{1}{2}(y_{k-1}^{(\nu+1)} - 2y_k^{(\nu)} + y_{k+1}^{(\nu)} - 0.4),$$

we now use the formula

$$y_k^{(\nu+1)} = y_k^{(\nu)} + 0.7(y_{k-1}^{(\nu+1)} - 2y_k^{(\nu)} + y_{k+1}^{(\nu)} - 0.4)$$

$$= 0.7y_{k-1}^{(\nu+1)} - 0.4y_k^{(\nu)} + 0.7y_{k+1}^{(\nu)} - 0.28.$$

In modification of (376), we then obtain the following table:

$x =$	0	1	2	3	4	5
$\nu = 1$	0	1	2	3	4	5
$\nu = 2$	0	0.72	1.524	2.3868	3.29076	5
$\nu = 3$	0	0.4988	1.13032	1.86004	3.20572	5

We observe the faster—as compared with (376)—convergence to the final values which, according to (373), are equal to:

$$0 \qquad 0.2 \qquad 0.8 \qquad 1.8 \qquad 3.2 \qquad 5$$

As was shown earlier, the linear system of equations (371) corresponding to our boundary-value problem, is symmetric definite. As a second available relaxation method we can therefore use the *method of steepest descent*, discussed in detail in 6.45. According to (346), it consists of integrating the system of differential equations

$$\frac{dy_k}{dt} = y_{k-1} - 2y_k + y_{k+1} - h^2 p_k \tag{379}$$

for the functions $y_k(t)$ by means of the Euler method. After selection of a step length l for the parameter t, the relaxation formula (347) has the form

$$y_k^{(\nu+1)} = y_k^{(\nu)} + l(y_{k-1}^{(\nu)} - 2y_k^{(\nu)} + y_{k+1}^{(\nu)} - h^2 p_k). \tag{380}$$

This is the method of *Richardson*; in contrast to the single-step method, it is called the *total-step method*. For the special choice $l = \frac{1}{2}$, we obtain the simple formula

$$y_k^{(\nu+1)} = \frac{y_{k-1}^{(\nu)} + y_k^{(\nu)}}{2} - \frac{h^2}{2} p_k$$

and, with the special numerical assumptions (372) and with $h = 1$, it then follows that

$$y_k^{(\nu+1)} = \frac{y_{k-1}^{(\nu)} + y_{k+1}^{(\nu)}}{2} - 0.2 \tag{381}$$

which can be compared with (375a). The numerical evaluation proceeds in a table similar to (376) whereby the index ν again numbers the rows:

0	1	2	3	4	5	
0	1	2	3	4	5	
0	0.8	1.8	2.8	3.8	5	(382)
0	0.7	1.6	2.6	3.7	5	

Here, for example, the values in the third row are formed by decreasing the mean value of the neighboring quantities in the 2nd row by 0.2. The procedure is fundamentally different from that of the single-step method. A value already found is *not* used in the computation of the following values of the same row.

The choice $l = \frac{1}{2}$ for the step in the t-variable requires some explanation. According to (349), this choice is decided by the largest eigenvalue λ_{max} of the eigenvalue problem that corresponds to the given equations (371). The homogeneous equations belonging to the equations (371) are

$$-y_{k-1} + 2y_k - y_{k+1} = 0, \qquad k = 1, 2, \ldots, (n-1)$$

where we have to set $y_0 = 0$ and $y_n = 0$. Hence, the eigenvalue problem reads

$$-y_{k-1} + 2y_k - y_{k+1} = \lambda y_k. \tag{383}$$

To each of these $(n-1)$ equations belongs a Gershgorin circle (5.24) in the complex λ-plane. All circles have the same center $\lambda = 2$. The first and the last ones have the radius 1, while all other circles have the radius 2. The domain covered by these circles is the circle which has the interval from 0 to 4 on the real λ-axis as diameter. Since the eigenvalue problem is symmetric, all eigenvalues are real and are therefore contained in the interval (0, 4). Hence,

$$\lambda_{max} \leq 4 \qquad \text{or} \qquad \frac{1}{2} \leq \frac{2}{\lambda_{max}}. \tag{384}$$

In our case, the condition (349) therefore assumes the form

$$l < \frac{2}{\lambda_{max}}.$$

Because of (384) it is satisfied if we choose $l < \frac{1}{2}$; $l = \frac{1}{2}$ is also admissible since a more precise investigation shows that the equality sign in (384) is not possible.[1] However, if the subdivision of the fundamental interval is refined more and more, λ_{max} increasingly approaches the value 4; as a result we are then, with $l = \frac{1}{2}$, close to the border of stability.

6.6 Partial Differential Equations

A systematic treatment of this important part of applied mathematics would exceed the limits set for this book. We must confine ourselves here to presenting some considerations which are of importance to numerical work and which tie in easily with our earlier discussions. As is well known, one distinguishes between partial differential equations of elliptic, parabolic, and hyperbolic types. Each of these categories has its own natural problems which differ according to the above-named types. In the case of elliptic equations, boundary-value problems generally have to be solved which describe

---

[1] For example, this follows from an explicit solution of the eigenvalue problem (383); compare 6.6, formulas (395) through (397).

configurations of physical equilibrium.   Parabolic equations are encountered with diffusion and heat-conduction problems while hyperbolic equations play a dominant role in the field of supersonic flows, in gas dynamics, and also appear in connection with propagation problems of all kinds of waves.   We shall not be concerned with this latter type.   Concerning the theory of hyperbolic equations, we refer to [24], [25], and for their numerical treatment to [21], [33].

6.61   Parabolic Equations

Consider the problem of finding a function $y(x, \tau)$ over the interval $(0, a)$ of the x-axis, which depends not only on x but also on the *time* τ, and which satisfies the partial differential equation

$$\frac{\partial y}{\partial \tau} = c\left[\frac{\partial^2 y}{\partial x^2} - p(x)\right]. \tag{385}$$

Here c is a given constant and $p(x)$ a given function.   Furthermore, the same *boundary conditions* are prescribed as with the ordinary boundary-value problem (368):

$$y(0) = 0, \qquad y(a) = b. \tag{386}$$

These conditions have to be satisfied at every time τ.   Finally, let the following *initial condition* for the course of the unknown function be given at $\tau = 0$:

$$y(x, 0) = f(x) \tag{387}$$

where $f(x)$ is known.   We therefore have a mixed problem here; boundary conditions are given for the spatial variable x, but an initial condition has been prescribed for the time variable τ.   The problem can be interpreted physically as heat conduction in a rod, covering the interval $(0, a)$ of the x-axis; $y(x, \tau)$ is the temperature which initially is equal to $f(x)$.   The ends of the rod are kept at the constant temperatures 0 and b respectively.

The discretization is carried out exactly according to the pattern used with the ordinary boundary-value problem (368); i.e., the fundamental interval is subdivided into n equal parts of the length h.   The partition points are then $x_k = kh$ and the corresponding I-values

$$y_k = y(x_k, \tau), \qquad p_k = p(x_k).$$

But the y_k are now functions of the time.   As before, we approximate the second derivative with respect to x by numerical differentiation as follows:

$$\frac{\partial^2 y}{\partial x^2}(x_k, \tau) \sim \frac{y_{k-1} - 2y_k + y_{k+1}}{h^2}.$$

Hence, the partial differential equation (385) can be replaced approximately by a system of ordinary differential equations

$$\frac{dy_k}{d\tau} = c\left[\frac{y_{k-1} - 2y_k + y_{k+1}}{h^2} - p_k\right], \qquad k = 1, 2, \ldots, (n-1). \quad (388)$$

Here we must remember that $y_0 = 0$, $y_n = b$. In order to use earlier results we introduce, instead of the time τ, the variable

$$t = \frac{c}{h^2}\,\tau \qquad (389)$$

thereby obtaining the system

$$\frac{dy_k}{dt} = y_{k-1} - 2y_k + y_{k+1} - h^2 p_k \qquad (390)$$

which is completely identical with (379). Let us look into this coincidence a little more closely. For this purpose, we investigate whether, after an infinitely long time, a *stationary state* of the temperature y is reached in our heat-conduction problem, i.e., a state which is no longer dependent on the time. According to (385), (386), and because of $\partial y/\partial \tau = 0$, this state $y(x)$ must satisfy the conditions

$$\frac{d^2 y}{dx^2} = p(x), \qquad y(0) = 0, \qquad y(a) = b. \qquad (391)$$

But this is exactly the boundary-value problem (368). *The method of steepest descent (379) solves this boundary value problem by determining the stationary state of the heat conduction problem* (385), (386). Since the systems (390) and (379) are identical, the method accomplishes this *by solving the heat-conduction problem by means of the difference method.* We can observe somewhat inaccurately, yet to the point, that the method of steepest descent for the treatment of a boundary-value problem consists in solving the heat-conduction problem connected with this boundary-value problem.

Having pointed out this relationship we now return to our parabolic problem (385)–(387). A *first method* for its solution is the exact integration of the system of differential equations (390) which is linear and has constant coefficients. Since the system is inhomogeneous, we need a particular solution $Y_k(t)$. This can be obtained from the stationary state, i.e., from the boundary-value problem (391). With our previous assumptions

$$p(x) = \text{const} = 0.4; \qquad a = 5, \qquad b = 5, \qquad h = 1, \qquad n = 5$$

a solution of the boundary-value problem, according to (373), is given by

$$Y(x) = \tfrac{1}{5}x^2, \qquad Y_k = \tfrac{1}{5}x_k^2.$$

With $h = 1$ it follows that

$$Y_{k-1} - 2Y_k + Y_{k+1} = \tfrac{1}{5}[x_{k-1}^2 - 2x_k^2 + x_{k+1}^2]$$
$$= \tfrac{1}{5}[(x_k - 1)^2 - 2x_k^2 + (x_k + 1)^2)] = \tfrac{2}{5} = 0.4.$$

Since the Y_k are independent of t, they do indeed form a solution of the system (390) and satisfy the given boundary conditions $Y_0 = 0$, $Y_5 = 5$. The differences $\eta_k = y_k - Y_k$ then must satisfy the homogeneous system of differential equations

$$\frac{d\eta_k}{dt} = \eta_{k-1} - 2\eta_k + \eta_{k+1} \tag{392}$$

whereby $\eta_0 = 0$; $\eta_n = \eta_5 = 0$. In addition, the initial condition (387), which has so far been ignored, has to be satisfied. After discretization it changes to

$$y_k\big|_{t=0} = f_k \tag{393}$$

where f_k is an abbreviation for $f(x_k)$. This initial condition is carried over to η_k as follows:

$$\eta_k\big|_{t=0} = f_k - Y_k. \tag{394}$$

With this, we have reduced the problem to integration of the homogeneous system (392), subject to the initial conditions (394). According to the methods elaborated in 6.43 starting with formula (320), this integration can be executed explicitly. It leads to the eigenvalue problem

$$\eta_{k-1} - 2\eta_k + \eta_{k+1} = -\lambda\eta_k, \qquad k = 1, 2, 3, 4 \tag{395}$$

for determination of the eigenvalues of the system of differential equations. Here we have to set $\eta_0 = 0$, $\eta_5 = 0$. This is nothing else but the eigenvalue problem of the system of springs which was used as basic illustration throughout Chapter 5. Incidentally, this eigenvalue problem can be solved explicitly—a fact which we deliberately suppressed in that chapter. The eigensolutions are

$$\eta_k = \sin\frac{m\pi}{5}x_k, \qquad m = 1, 2, 3, 4. \tag{396}$$

Indeed, it follows that $\eta_0 = 0$, $\eta_5 = 0$ and

$$\eta_{k-1} - 2\eta_k + \eta_{k+1} = \sin\frac{m\pi}{5}(x_k - 1) - 2\sin\frac{m\pi}{5}x_k + \sin\frac{m\pi}{5}(x_k + 1)$$

$$= 2\left(\sin\frac{m\pi}{5}x_k\right)\left[\cos\frac{m\pi}{5} - 1\right] = -4\left(\sin^2\frac{m\pi}{10}\right)\eta_k$$

The eigenvalues are therefore

$$\lambda_m = 4 \sin^2 \frac{m\pi}{10}, \qquad m = 1, 2, 3, 4. \tag{397}$$

We denote the eigensolutions (396) more precisely by η_{km}. According to (322) and (322a), the solutions of the homogeneous system of differential equations are then given by

$$\eta_k = \sum_{m=1}^{4} \alpha_m \eta_{km} e^{-\lambda_m t} \qquad \text{where} \qquad \alpha_m = \frac{\sum_{k=1}^{4} (f_k - Y_k) \eta_{km}}{\sum_{k=1}^{4} \eta_{km}^2}. \tag{398}$$

Finally, the solution of the heat-conduction problem is obtained by adding the particular solution Y_k:

$$y_k = Y_k + \sum_{m=1}^{4} \alpha_m \eta_{km} e^{-\lambda_m t}. \tag{399}$$

This formula shows that with increasing time the temperature distribution does indeed tend to the stationary distribution Y_k. Whether this first (and very accurate) method can be carried over to general cases depends essentially on whether the corresponding eigenvalue problem lends itself to numerical treatment or not.

It has also been suggested that the system of differential equations resulting from the discretization should be solved by means of analogue computers. Only larger computers are capable of this task since as many electronic integrators are needed as partition points have been chosen in the basic interval. For sizable diffusion problems this requirement often results in a prohibitively large number of integrators. In that case, little else remains to be done but to return repentantly to the digital computer and to apply one of the following methods.

The *second method* consists of integrating the system of differential equations (390) or (379) by means of the Euler method. In other words, the time t is now also discretized, using a step length l. This is equivalent to the total step method for which we wrote out the computational procedure for $l = \frac{1}{2}$ in formula (381). If, in particular, we choose for the initial condition (387):

$$f(x) = x, \qquad \text{or} \qquad f_k = x_k,$$

then the calculation (or at least its beginning) is contained in the table (382). In it the time increases vertically downward. Each row contains the temperature distribution at a given moment; from one row to the next, t increases by $l = \frac{1}{2}$. In terms of the original time variable τ this step is equivalent to the increment

$$\Delta\tau = \frac{h^2}{2c}$$

as can be seen from (389). Following table (382) we discussed the selection of the step length and showed that the step l must not be chosen any larger if instabilities are to be avoided. Between the spatial step $\Delta x = h$ and the time step $\Delta \tau$ the relation

$$\Delta \tau \leq \frac{\Delta x^2}{2c} \tag{400}$$

ought therefore to be maintained. With some problems this condition causes such a fine subdivision of time that an intolerable amount of computational labor results. However, it is obvious that this second method will work very well for *small times*, while the first method is better suited for *large times*. In fact, for large t the exponential terms in (399), corresponding to large eigenvalues λ_m, have probably decreased sufficiently so that only a few λ_m need to be calculated at the lower end of the spectrum of the λ_m. More precisely, one proceeds as follows: Let the eigenvalues be arranged in order of increasing magnitude:

$$\lambda_1 < \lambda_2 < \lambda_3 < \lambda_4.$$

Now assume, for example, that we want to disregard λ_3 and λ_4. We therefore calculate only λ_1, λ_2 and the corresponding eigensolutions

$$\eta_{11}, \eta_{21}, \eta_{31}, \eta_{41} \quad \text{and} \quad \eta_{12}, \eta_{22}, \eta_{32}, \eta_{42},$$

respectively. From (398) we obtain

$$\alpha_1 = \frac{\sum_{k=1}^{4} (f_k - Y_k)\eta_{k1}}{\sum_{k=1}^{4} \eta_{k1}^2}, \qquad \alpha_2 = \frac{\sum_{k=1}^{4} (f_k - Y_k)\eta_{k2}}{\sum_{k=1}^{4} \eta_{k2}^2}$$

and from (399)

$$y_k \cong Y_k + \alpha_1 \eta_{k1} e^{-\lambda_1 t} + \alpha_2 \eta_{k2} e^{-\lambda_2 t}, \qquad k = 1, 2, 3, 4.$$

Here the symbol $\cong$ denotes asymptotic approximation, i.e., an approximation which improves more and more for increasing t. For small t this approximation is of course unusable.

The condition (400) for the step in the time direction can be avoided by choosing an integration procedure for the system (390) which is always stable. For this the integration with the trapezoidal rule recommends itself. As we saw in our discussion following the equation (334), this method has unrestricted stability. The trapezoidal integration of (390) over the time interval from t to $(t + l)$ yields

$$y_k(t + l) - y_k(t)$$

$$= \frac{l}{2} [y_{k-1}(t) - 2y_k(t) + y_{k+1}(t) + y_{k-1}(t + l) - 2y_k(t + l) + y_{k+1}(t + l)]$$

$$- lh^2 p_k.$$

For given $y_k(t)$ this is a linear system of equations for the values $y_k(t + l)$ of the next time step and this system is to be solved subject to the conditions

$$y_0(t + l) = 0, \qquad y_n(t + l) = b.$$

The matrix of this system is exactly the triple diagonal matrix of the system (369) which was written down in detail in example (370). Roughly speaking, at each temporal integration step we must therefore solve a boundary-value problem of the type (368) in its form (369) resulting from discretization. However, the computational work involved here is not too bad. The matrix of the system of equations is the same for all times t and therefore the Gauss algorithm has to be repeated only in the column of the constants.

We should also keep in mind the method of the *improved trapezoidal rule* [6.3, the part following the numerical example, i.e., after (287)]. In conclusion we should like to add that restricting stability conditions for the length of the time step will also have to be observed when integrating the system of differential equations (390) by means of the *Runge-Kutta* methods.

6.62 Elliptic Boundary-Value Problems

Let the rectangle shown in Fig. 28 be given as fundamental domain G in

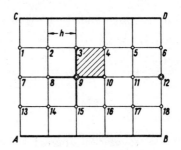

FIG. 28.   Elliptic boundary-value problem.

the x, y-plane. A function $u(x, y)$ is to be computed in G, which first of all satisfies the *Poisson*-differential equation

$$u_{xx} + u_{yy} = p(x, y). \tag{401}$$

Here $p(x, y)$ is a function in G and the indices denote partial derivatives

$$u_{xx} = \frac{\partial^2 u}{\partial x^2}, \qquad u_{yy} = \frac{\partial^2 u}{\partial y^2}.$$

Secondly, the following boundary conditions are prescribed on the horizontal sides:

$$\text{on} \quad AB \quad \text{and} \quad CD: \qquad u = 0. \tag{402}$$

In the third place, on the vertical sides of the rectangle the given boundary conditions are:

$$\text{on} \quad AC \quad \text{and} \quad BD: \qquad u_x = \frac{\partial u}{\partial x} = 0. \tag{403}$$

The problem can be interpreted physically as the determination of the equilibrium position of a membrane (rubber sheet) which—in the absence of outside forces—covers the rectangle G. The horizontal sides are nailed down, but the vertical sides are free. Now the membrane is subjected to a continuous force with the density $p(x, y)$ and this force acts vertically on the plane of our drawing. The desired function $u(x, y)$ then is the displacement of the membrane.

For the discretization of this problem, we have placed in Fig. 28 a quadratic net of mesh width h into the fundamental domain G.[1] At the 18 marked and numbered grid points the values $u_1, u_2, \ldots, u_{18}$ of the function $u(x, y)$ shall now be calculated. The grid points on the horizontal borders do not have to be introduced, since there the function u is known from the boundary conditions (402). In order to execute the discretization, we go first to one of the grid points in the interior of G—for example, to point $\#9$. The numerical differentiation there yields approximately

$$u_{xx} = \frac{u_8 - 2u_9 + u_{10}}{h^2}, \qquad u_{yy} = \frac{u_3 - 2u_9 + u_{15}}{h^2}.$$

From (401) we then obtain the equation

$$\frac{u_8 - 2u_9 + u_{10}}{h^2} + \frac{u_3 - 2u_9 + u_{15}}{h^2} = p_9$$

where p_9 is the value of $p(x, y)$ at the grid point $\#9$. This equation can also be written in the form

$$4u_9 - u_{10} - u_3 - u_8 - u_{15} + h^2 p_9 = 0 \tag{404}$$

and in Fig. 29 it is represented in a symbolic form applicable to each of the 12 interior grid points. The cross in this figure is called the *operator* of the problem; at its points, the coefficients of the linear equation (404) have been written down. It should be added that, for example, for the operator with the center at grid point $\#4$, the upward-pointing arm of the cross will have to be omitted since the value of $u(x, y)$ at its end point is equal to zero, due

---

[1] For simplicity's sake, we assume that this is possible, i.e., that the sides of the rectangle are integral multiples of h. Later on, we shall take up the general case where the boundary of G does not coincide with the grid lines of the quadratic grid, or is even curved.

to boundary condition (402). Corresponding to the 12 interior grid points, we obtain 12 linear equations for the 18 unknowns $u_1, u_2, \ldots, u_{18}$.

The 6 missing equations are provided by the boundary conditions (403). The first formula of Appendix II, Table 1, for differentiation at the beginning gives for example at boundary point #1

$$\frac{\partial u}{\partial x} = \frac{-3u_1 + 4u_2 - u_3}{2h}.$$

From (403) therefore the equation

$$-3u_1 + 4u_2 - u_3 = 0 \tag{405}$$

follows. Consequently, we have found $12 + 6 = 18$ equations for our 18 unknowns and with this the algebraization of our problem is concluded.

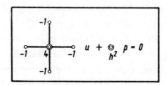

FIG. 29.   Operator equation for interior points.

However, this *first method* of discretization is too crude for our present problem. The *second method*, which follows, is more advantageous; it is also called the *energy method*. The deformation energy of the membrane is

$$E_1 = \tfrac{1}{2} \iint\limits_{(G)} (u_x^2 + u_y^2)\, dx\, dy \tag{406}$$

and the work of the forces applied is

$$E_2 = \iint\limits_{(G)} p(x, y) u(x, y)\, dx\, dy. \tag{407}$$

According to the variational principles of mechanics, the function $u(x, y)$ can therefore also be characterized by the fact that it minimizes the expression

$$E = E_1 + E_2 = \tfrac{1}{2} \iint\limits_{(G)} (u_x^2 + u_y^2)\, dx\, dy + \iint\limits_{(G)} pu\, dx\, dy. \tag{408}$$

To this the boundary conditions (402) must be added, while *the boundary conditions (403) need not be considered at all.* In fact, with the help of the calculus of variations it is possible to show that the minimal function u satisfies the conditions (403) *automatically*.   This reduction of the boundary conditions is one of the *first advantages* of the energy method.

The new method of discretization consists of approximating the energy expressions E_1, E_2 by means of the difference method.   Only after this discretization do we proceed with the minimization.   In order to approximate first E_1, we calculate approximately the energy contained in one mesh square of our grid.   For example, let us consider the shaded mesh square in Fig. 28.   According to the rule of numerical differentiation (6.1), we have at the center points of the grid lines 3, 4 and 9, 10, the approximations

$$u_x = \frac{u_4 - u_3}{h} \quad \text{and} \quad u_x = \frac{u_{10} - u_9}{h},$$

respectively. The arithmetic mean of the squares of these expressions is a good approximation for the value of u_x^2 in the center of the mesh square:

$$u_x^2 = \frac{1}{2h^2}\left[(u_4 - u_3)^2 + (u_{10} - u_9)^2\right].$$

Similarly, it follows that at the same center point

$$u_y^2 = \frac{1}{2h^2}\left[(u_4 - u_{10})^2 + (u_3 - u_9)^2\right].$$

Since the mesh square has the area h^2, its energy content according to (406) is therefore approximately equal to

$$E_1: \quad \tfrac{1}{4}[(u_4 - u_3)^2 + (u_{10} - u_9)^2 + (u_4 - u_{10})^2 + (u_3 - u_9)^2]. \quad (409)$$

The content of each parenthesis can be attributed to one grid line on the boundary of our mesh square.   We can therefore consider the grid lines as the carriers of the energy E_1.   However, we will not use this interpretation since it is linked too closely to the special nature of our problem.   Now E_1 is approximately the sum of all these contributions of the individual mesh squares.   For the contribution of the same shaded mesh square to the energy E_2, we use the approximate expression

$$E_2: \quad \frac{h^2}{4}\left[p_3 u_3 + p_4 u_4 + p_9 u_9 + p_{10} u_{10}\right]. \quad (410)$$

Apart from the factor h^2, we have here the arithmetic mean of the values of the integrand of (407) at the four corners.   This formula therefore represents a kind of two-dimensional trapezoidal rule.

We now recognize that the expression $E = E_1 + E_2$ to be minimized is a quadratic function of the variables $u_1, u_2, \ldots, u_{18}$. This function is built up out of the quadratic form E_1 and the linear function E_2. Equation (409) shows that—as a sum of squares—the form E_1 is positive definite. Our problem therefore turns out to be a *definite problem*. Consequently, the linear system of equations resulting from minimization of E is *symmetric definite*. This is the *second advantage* of the energy method. On the one hand, because of the symmetry, the computational labor for execution of the Gauss algorithm has been reduced by almost one half; on the other hand, the definiteness makes the application of relaxation techniques possible.

We now proceed to the problem of setting up this linear system of equations. E is to be minimized, hence E has to be differentiated with respect to every one of the variables $u_1, u_2, \ldots, u_{18}$. The calculation will be slightly different, depending on whether the variable under consideration corresponds to a grid point within the rectangle or on its border. We begin with interior point #9. The corresponding variable u_9 occurs only in the contributions of those mesh squares which have point #9 as one of their corners. We therefore need to consider only the four mesh squares which meet at the point #9. After differentiation of its E_1 contribution (409) with respect to u_9, the square shaded in Fig. 28 yields:

$$\tfrac{1}{2}(-u_3 + 2u_9 - u_{10}). \tag{411}$$

The remaining three squares give the similar contributions

$$\tfrac{1}{2}(-u_3 + 2u_9 - u_8), \qquad \tfrac{1}{2}(-u_8 + 2u_9 - u_{15}),$$
$$\tfrac{1}{2}(-u_{15} + 2u_9 - u_{10}). \tag{412}$$

Combining these four partial results we obtain

$$\frac{\partial E_1}{\partial u_9} = 4u_9 - u_{10} - u_3 - u_8 - u_{15}.$$

Differentiation of the E_2-contribution (410) of the shaded mesh square with respect to u_9 leads to the partial result

$$\frac{h^2}{4} p_9$$

and the four squares together yield

$$\frac{\partial E_2}{\partial u_9} = h^2 p_9. \tag{413}$$

Hence, to the interior grid point #9 belongs the equation

$$\frac{\partial E_1}{\partial u_9} + \frac{\partial E_2}{\partial u_9} = 0 \qquad \text{or} \qquad 4u_9 - u_{10} - u_3 - u_8 - u_{15} + h^2 p_9 = 0. \tag{414}$$

It is completely identical with equation (404) which was found with the help of the first discretization method.   So far, therefore, the energy method has not given us anything new.   But it is a different story where a point on the boundary is concerned.   As an example we use point #12.   In it only two squares of the grid meet; they generate contributions similar to (412):

$$\tfrac{1}{2}(-u_6 + 2u_{12} - u_{11}), \qquad \tfrac{1}{2}(-u_{11} + 2u_{12} - u_{18}).$$

Hence,

$$\frac{\partial E_1}{\partial u_{12}} = \tfrac{1}{2}(4u_{12} - 2u_{11} - u_6 - u_{18})$$

and analogous to (413) we have for E_2

$$\frac{\partial E_2}{\partial u_{12}} = \frac{h^2}{2}\, p_{12}.$$

To point #12 therefore belongs the equation

$$4u_{12} - 2u_{11} - u_6 - u_{18} + h^2 p_{12} = 0 \qquad (415)$$

which is given in operator form in Fig. 30.   This operator equation applies to every point on the two sides AC and BD; on AC the horizontal leg of the

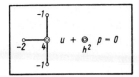

FIG. 30.   Operator equation for the boundary points.

"half"-cross points, of course, to the right.   The equation is entirely different from equation (405) which was found by means of the first method; it contains implicitly the correct consideration of the boundary condition (403) which is coherent with the other approximations.   The 12 equations of the type in Fig. 29 and the 6 equations of the type in Fig. 30 now constitute our symmetric definite system of equations.   Let us add a word concerning its solution by means of a *relaxation method.*

The *single-step method* is again obtained when we solve the equations of Figs. 29 and 30 with respect to the central unknown, i.e., the unknown belonging to the center of the cross.   At the first grid points #1, 2, we then have

$$u_1 = \frac{2u_2 + u_7 - h^2 p_1}{4}, \qquad u_2 = \frac{u_1 + u_3 + u_8 - h^2 p_2}{4}. \qquad (416)$$

We now choose initial approximations for the u; through substitution in the left-hand formula we determine from them an improved value of u_1, which is immediately used in the right-hand formula. This in turn yields an improved value of u_2. In this way we proceed through the entire field in the same sequence as the lines of a book are read—after which we start again in the left-hand upper corner.

For example, consider the *heat-conduction problem* for a function $u(x, y; \tau)$:

$$\frac{\partial u}{\partial \tau} = u_{xx} + u_{yy} - p(x, y)$$

with the boundary conditions (402), (403), and certain initial conditions for $\tau = 0$. Discretization is performed with exactly the same energy method. Instead of $p(x, y)$ in (402) we need only write the expression

$$\frac{\partial u}{\partial \tau} + p(x, y).$$

Instead of (416), we then obtain the differential equations

$$\frac{du_1}{d\tau} = \frac{-4u_1 + 2u_2 + u_7}{h^2} - p_1, \qquad \frac{du_2}{d\tau} = \frac{-4u_2 + u_1 + u_3 + u_8}{h^2} - p_2.$$

Altogether we find a system of 18 simultaneous differential equations. Integrating them by means of Euler's method then is equivalent to using the *total-step method* for the solution of the original boundary-value problem (401) through (403). The time step $\Delta\tau$ is subject to the condition

$$\Delta\tau \leq \frac{h^2}{4}$$

which is based on similar considerations, as is the condition (400).

Electronic computers must be used for single- as well as total-step computations and the earlier mentioned convergence-acceleration techniques of Young, Flanders-Shortly, Hestenes-Stiefel have to be applied. All these refined methods are collected in [16]. There a description can also be found of the energy method for simple elasticity problems; [33] contains a thorough and modern description of the difference methods for all types of partial differential equations.

6.63 Curved Boundary

The minimization problem (408) is now to be solved in a fundamental domain G which may have a curved boundary. Again we use a quadratic grid with the mesh width h which covers G. A first and simple, but very rough, method of discretization consists in constructing the *grid domain* G' consisting of all mesh squares which are completely contained in G (Fig. 31). Its boundary consists of grid lines, and because of its staircase-like appearance, it is also called the *staircase boundary*. We introduce as unknowns u_k the values of $u(x, y)$ at the grid points contained in G' or on the staircase boundary (in Fig. 31 these are indicated by little blank circles). The approximation

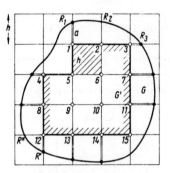

FIG. 31. Curved boundary.

for the energies E_1, E_2 is obtained exactly according to our previous methods and formulas. All mesh squares of the grid domain G' must be considered. The contribution to energy E_1 of the shaded mesh square 1, 2, 5, 6 would, for example, be given by

$$E_1: \quad \tfrac{1}{4}[u_1 - u_5)^2 + (u_5 - u_6)^2 + (u_6 - u_2)^2 + (u_2 - u_1)^2]. \quad (417)$$

We still have to discuss the treatment of a possible boundary condition given on the boundary of the fundamental domain G. For example, let the condition $u(x, y) = 0$ be prescribed on this boundary. We shall only satisfy this condition approximately in certain boundary points of G located on grid lines. For example, let the point R_1 be such a boundary point in Fig. 31. Proceeding along its grid line we come to point #1 of the grid domain and then to point #5. If the value of u at R_1 is denoted by $u(R_1)$ for the time being, then linear interpolation using the points on the grid line yields

$$u_1 = \frac{au_5 + hu(R_1)}{a + h}$$

and since $u(R_1)$ shall be zero:

$$u_1 = \frac{a}{a + h} u_5.$$

With the help of this equation it is then possible to eliminate the variable u_1 from the entire problem. We therefore call point #1 of the staircase boundary an *elimination point* and point #5 the corresponding *auxiliary point*. As a result of this elimination the energy expression (417) assumes the changed form

$$E_1: \quad \frac{1}{4}\left[\frac{h^2}{(a+h)^2}u_5^2 + (u_5-u_6)^2 + (u_6-u_2)^2 + \left(u_2 - \frac{a}{a+h}u_5\right)^2\right],$$

(418)

thus becoming a quadratic form in the variables u_2, u_5, u_6. When choosing further boundary points R_i we must see to it that the needed auxiliary points are not former elimination points. For example, the boundary point R_2 was not chosen for the elimination since otherwise the grid point #2 would be an elimination point, and this same point occurs as auxiliary point

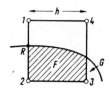

Fig. 32.   Curved boundary.

in connection with R_3. Furthermore, it is clear that no two boundary points may be chosen which lead to the same elimination point (for example, the points R', R'' in the area of corner #12). After all eliminations have been introduced in the energy expressions we again differentiate with respect to the remaining variables in order to obtain the operator equations.

A more refined approximation method consists in also using all those mesh squares which are cut by the boundary of G. Figure 32 shows one of these squares with the corner points numbered 1, 2, 3, 4. The corresponding u-values u_1, u_2, u_3, u_4 appear as variables in our problem, although u_1, u_4 belong to grid points outside of the fundamental domain. Let F be the area of that portion of the fundamental domain which is contained in the square. As contribution of the square to E_1, we can then choose the expression

$$E_1: \quad \frac{F}{4h^2}[(u_1-u_2)^2 + (u_2-u_3)^2 + (u_3-u_4)^2 + (u_4-u_1)^2].$$

Here the factor (F/h^2) takes care of the reduction of the area involved. Boundary conditions are again satisfied through elimination. In Fig. 32, for example, we can choose the boundary point R and associate with it the elimination point #1 and the auxiliary point #2.

More complicated boundary conditions usually require that a linear combination of certain grid values u_i equal a given value. Such conditions can be introduced as subsidiary conditions to the minimization problem and can be handled by means of the method of *Lagrange multipliers*. Another way would be to modify the methods of steepest descent in such a way that the subsidiary conditions are satisfied at each step. If a problem does not have definite character, the energy method is no longer applicable, and only the first method of discretization remains. In that case, directions can be found in [21], [33] for the treatment of curved boundaries.

7 Approximations

In this chapter we shall be concerned with the approximate representation of a function $f(x)$ given by I-values as, for example, in the form of a table. An ordinary or trigonometric polynomial is to be constructed which approximates these I-values well. First, the I-polynomial will be constructed which passes exactly through these I-values; later, approximations by means of polynomials of lower degree are to be investigated whereby the Chebyshev polynomials will play an important role. The entire problem area is connected with interpolation, but interpolation itself is not the actual target of our investigations. Some concepts and results of Chapter 6 will be needed.

It is often more economical to use rational functions instead of polynomials for the approximation. In this connection we refer to [32].

7.1 The Interpolating Polynomial

As in 6.1, let us assume that $(n+1)$ pairwise different I-places $x_0, x_1, \ldots, x_n$ are given on the x-axis, and that they are numbered in increasing order:

$$x_0 < x_1 < \ldots < x_n.$$

They are not necessarily equidistant. Before dealing with any approximation problem it is advisable to assign an *interpolating coefficient* λ_i to every I-place x_i. This coefficient is defined as follows:

$$\lambda_i = \frac{1}{(x_i - x_0) \ldots (x_i - x_{i-1})(x_i - x_{i+1}) \ldots (x_i - x_n)}. \tag{419}$$

The denominator contains the product of the distances from the particular I-place x_i to all other I-places. When numbered in the order $i = 0, 1, \ldots, n$, the I-coefficients have alternating signs; λ_n is positive. For practical computation of the λ_i it is best to prepare a distance table similar to those found on maps or time tables.

Example. Let 0, 1, 3, 4 be the given I-places.

		x_0	x_1	x_2	x_3		
		0	1	3	4		
x_0	0		1	3	4	12	
x_1	1	1		2	3	6	
x_2	3	3	2		1	6	(420)
x_3	4	4	3	1		12	
λ		-1	2	-2	1	12	

The distances are marked down without concern for the signs. The product of the numbers in a row is placed into the last column. Next, the reciprocal values are placed in the lowest row, and finally, the correct signs are inserted. In the example only the numerators of the λ_i are written down, the common denominator 12 stands in the lower right-hand corner. In more complicated cases it is advantageous to use logarithms for the computation.

In the case of equidistant I-places, a simple formula can be found for the I-coefficients. Let the I-places be the integers

$$0, 1, 2, \ldots, \rho, \ldots, n.$$

Then the I-place ρ has the distances $1, 2, \ldots, \rho$ from its left-hand neighbors and the distances $1, 2, \ldots, (n - \rho)$ from its right-hand neighbors. Hence,

$$|\lambda_\rho| = \frac{1}{\rho! \, (n - \rho)!} = \frac{1}{n!} \binom{n}{\rho}. \tag{421}$$

The signs of the λ_ρ again alternate. If the I-places follow each other with a constant step length h, we have the formula

$$|\lambda_\rho| = \frac{1}{h^n n!} \binom{n}{\rho}. \tag{422}$$

Except for a constant factor the λ_ρ are therefore the alternating binomial coefficients.

If one abscissa has been left out in a system $x_0, x_1, \ldots, x_n$ of I-places we shall call the system of the remaining n places a *reference system* of the nth level. To this reference system again belong n I-coefficients which are generally completely different from those of the entire system. When two I-places are left out we obtain a reference system of the $(n - 1)$th level, and so on. These reference systems and their I-coefficients form an important tool in the solution of approximation problems. In Appendix II, Table 3, reference systems of all levels have been chosen for 5 and 9 equidistant I-places, and all corresponding I-coefficients have been calculated. The selection of the reference systems was done in a way which is particularly advantageous to approximation problems, but we cannot enter into a discussion of the reasons for this here.[1] However, we note the fact that in a fixed reference system the I-places are distributed more densely near the ends of the I-interval than near the middle. This has a numerically stabilizing effect (compare also the following discussion of the Lagrange formula).

Let us return to a system $x_0, x_1, \ldots, x_n$ of I-places, and assume that the corresponding I-values $y_0, y_1, \ldots, y_n$ are also given. The I-polynomial $P_n(x)$ of nth degree is now to be constructed which passes through these I-values. Both its existence and its uniqueness have already been proved in 6.1. The following Lagrange formula solves this I-problem at one sweep. We have

$$P_n(x) = \sum_{i=0}^{n} \lambda_i (x - x_0) \ldots (x - x_{i-1})(x - x_{i+1}) \ldots (x - x_n)y_i. \quad (423)$$

In addition to λ_i, y_i, the ith summand contains the product of the distances from the point x to all I-places with the exception of the ith one. The formula is easily verified. First of all, the ith summand is a polynomial of nth degree in x, as can be seen after expanding the product of the distances. The sum $P_n(x)$ is therefore a polynomial of nth degree as well. In the second place, if we now go to any one of the I-places, i.e., if we set $x = x_j$, all summands with the exception of the jth one disappear. Consequently, we have

$$P_n(x_j) = \lambda_j(x_j - x_0) \ldots (x_j - x_{j-1})(x_j - x_{j+1}) \ldots (x_j - x_n)\,y_j = y_j$$

where we used the definition (419) of the I-coefficient λ_j. $P_n(x)$ therefore does indeed assume all the I-values. In the case of equidistant I-places, the *Lagrange coefficients*

$$l_i(x) = \lambda_i(x - x_0) \ldots (x - x_{i-1})(x - x_{i+1}) \ldots (x - x_n) \quad (424)$$

needed in (423), have been tabulated in detail as functions of x [27]. The use of such a table reduces the calculation of the I-polynomial to the application of the formula

$$P_n(x) = \sum_{i=0}^{n} l_i(x)y_i. \quad (425)$$

---

[1] The I-places in the reference system of the ρth level are the extremal points of the so-called S-function $S_{\rho-1}$ (compare [26]).

This is a linear combination of the I-values y_i. A look into such a table is very instructive. For example, let us take 11 equidistant I-places, in particular the integers from (-5) to $(+5)$ (Fig. 33). As long as x is contained in the interval $(-3, +3)$, the absolute values of the Lagrange coefficients l_i exceed the value 1 only insignificantly. Outside of this interval, however, some of these l_i assume considerably larger values. For example, for $x = 4.7$ the coefficient corresponding to the central I-abscissa $x_5 = 0$ is equal to 6. This means that any inaccuracy of the I-value y_5

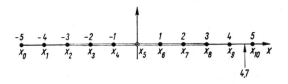

FIG. 33. To the Lagrange formula.

belonging to this center point is transferred to the value of the I-polynomial $P_{10}(4.7)$ with 6-fold magnification. This in turn is almost equal to the loss of one decimal place. The I-polynomial therefore is accurately determined only in the center region of the I-interval; toward the ends of the interval, its course becomes uncertain. This situation grows rapidly worse with an increase in the number of I-places. Consequently, there is not much point in trying to connect a large number of measured equidistant I-values by means of the I-polynomial. The inaccuracy at the ends is increased even further when the *Newton interpolation formula* (compare 7.12) is applied without any proper precautionary measures. If, in place of the equidistant abscissas, I-places are used which are closer together near the ends of the interval, then this "flutter" of the I-polynomial can be avoided. The *Chebyshev abscissas* discussed later on are an example of a simple and safe distribution of the I-places.

7.11   Computational Techniques for the Lagrange Formula

We call *interpolation* the problem of computing the value of the interpolating polynomial at a single numerically given point x. This point shall be called the "new point", or N-point for short. In order to simplify the evaluation of (423) we expand the distance product by the additional factor $(x - x_i)$:

$$P_n(x) = \sum_{i=0}^{n} \frac{\lambda_i}{(x - x_i)} \left[(x - x_0)(x - x_1) \ldots (x - x_n) \right] y_i.$$

The term in the square brackets is independent of the summation index i and can therefore be placed before the sum. With the abbreviations

$$\mu_i = \frac{\lambda_i}{x - x_i}, \qquad Q_n(x) = (x - x_0)\,(x - x_1)\,...\,(x - x_n) \qquad (426)$$

we then find

$$P_n(x) = Q_n(x) \sum_{i=0}^{n} \mu_i y_i. \qquad (427)$$

To simplify this further, we consider the special case where all I-values are equal to 1. The I-polynomial $P_n(x)$ is then the constant 1 and (427) reduces to

$$1 = Q_n(x) \sum \mu_i, \qquad \text{i.e.,} \qquad Q_n(x) = \frac{1}{\sum \mu_i}.$$

With this the Lagrange formula (427) assumes its *interpolatory form*:

$$\boxed{P_n(x) = \frac{\sum \mu_i y_i}{\sum \mu_i}, \qquad \text{where} \qquad \mu_i = \frac{\lambda_i}{x - x_i}} \qquad (428)$$

This shows that at the N-point the ordinate of the I-polynomial is equal to the weighted mean of the I-values with the numbers μ_i as weights. Note that instead of the μ_i we can also use weights which are proportional to the μ_i. Note that not all the weights are positive. In fact, by using, if necessary, the weights $(-\mu_i)$ instead of the weights μ_i, we find from the sign rules for the λ_i that the following is true: The two I-places immediately adjacent to the N-point have positive weights and the weights of all other I-places farther to the left and right have alternating signs.

When the N-point is very close to one of the I-places, say to x_j, then the weight μ_j becomes very large, a fact which could be very inconvenient (at least for automatic computation). In this case, the weights

$$\mu_i' = \lambda_i \frac{x - x_j}{x - x_i}$$

can be used throughout, whereby x_j is the I-place closest to the N-point.

1. **Example.** Consider the case presented in table (420) (I-places 0, 1, 3, 4). The value of the I-polynomial $P_3(x)$ is to be calculated at the N-point $x = 2$. The common denominator 12 of the λ_i can be disregarded:

$$\mu_0 = -\frac{1}{2}, \qquad \mu_1 = \frac{2}{1}, \qquad \mu_2 = \frac{2}{1}, \qquad \mu_3 = -\frac{1}{2}$$

$$P_3(2) = \frac{1}{6}\,(-y_0 + 4y_1 + 4y_2 - y_3).$$

2. Example. *Two I-places x_0, x_1 a distance h apart.* (Linear interpolation).
According to (419), we have

$$\lambda_0 = -\frac{1}{h}, \qquad \lambda_1 = \frac{1}{h}.$$

We introduce the *relative distances* between the N-point and the I-places:

$$p = \frac{x - x_0}{h}, \qquad q = \frac{x_1 - x}{h}, \qquad p + q = 1. \tag{429}$$

Then it follows from (424) that

$$l_0(x) = q, \qquad l_1(x) = p$$

and the original Lagrange formula (425) yields the well-known result

$$P_1(x) = qy_0 + py_1 \tag{430}$$

for the *linear interpolation* in the form best suited to computation with desk
calculators.

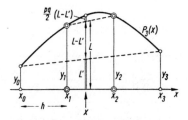

FIG. 34. Cubic interpolation.

3. Example. *Four equidistant I-places x_0, x_1, x_2, x_3.* (Step h) (Fig. 34). Let
the N-point be contained in the center interval (x_1, x_2). We again introduce
the relative distances:

$$p = \frac{x - x_1}{h}, \qquad q = \frac{x_2 - x}{h}, \qquad p + q = 1. \tag{431}$$

Except for a proportionality factor the distances from the N-point x to the
I-places are then equal to:

$$1 + p, \qquad p, \qquad q, \qquad 1 + q.$$

Again, except for a proportionality factor, the absolute values of the I-co-
efficients are equal, according to (422), to the binomial coefficients 1, 3, 3, 1.
We can therefore use the following values

$$-\frac{1}{1 + p}, \qquad \frac{3}{p}, \qquad \frac{3}{q}, \qquad -\frac{1}{1 + q}$$

for the weights μ_i. Here the signs were inserted according to the above-stated rule. After multiplication by the common denominator $pq(1 + p)$ $(1 + q)$, we arrive at the set of weights:

$$-pq(1 + q), \quad 3q(1 + p)(1 + q), \quad 3p(1 + p)(1 + q), \quad -pq(1 + p)$$

or, because of $p + q = 1$:

$$-pq(1 + q), \quad 3q(2 + pq), \quad 3p(2 + pq), \quad -pq(1 + p).$$

The sums of these weights equals

$$-pq(2 + p + q) + 3(2 + pq)(p + q) = -3pq + 3(2 + pq) = 6.$$

According to (428), the desired interpolation formula then assumes the form

$$P_3(x) = \tfrac{1}{6}[-pq(1 + q)y_0 + 3q(2 + pq)y_1 + 3p(2 + pq)y_2 - pq(1 + p)y_3].$$

We can also give it the following more convenient form. Let L be the result obtained by linear interpolation over the center interval, i.e., according to (430),

$$L = qy_1 + py_2.$$

Then

$$\boxed{P_3(x) = L + \frac{pq}{2}\left[L - \frac{(1 + q)y_0 + (1 + p)y_3}{3}\right]} \tag{432}$$

The fraction in the square brackets is simply the result of linear interpolation over the large interval (x_0, x_3) denoted in Fig. 34 by L'. Cubic interpolation can therefore be generated in a simple way from two linear interpolations.

In order to calculate the second summand in (432) we need to evaluate an expression of the form

$$B = \frac{pq}{2}A. \tag{433}$$

This can be facilitated by use of the nomogram in Appendix II, Table 8.

In order to give a numerical example, we take the following values from a logarithm table:

x	2	2.5	3	3.5	
$\log x$	0.3010	0.3979	0.4771	0.5441	(434)

The value log 2.63 is to be calculated approximately by interpolation.

$$p = 0.26, \quad q = 0.74, \quad \frac{pq}{2} = 0.0962.$$

Formula (432) yields:

$$
\begin{array}{ll}
L = 0.41849 & \quad L = 0.41849 \\
\underline{L' = 0.40310} & \quad \underline{148} \\
L - L' = 0.01539 & \quad 0.41997, \qquad \log 2.63 \sim 0.4200
\end{array}
\tag{435}
$$

Coincidentally, the result is accurate to four decimal places.

7.12 Application to Numerical Differentiation and Interpolation

From the Lagrange formula (423) we immediately obtain the following value for the highest coefficient of the interpolating polynomial (coefficient of x^n):

$$
\text{Highest coefficient of } P_n(x) = \sum_{i=0}^{n} \lambda_i y_i. \tag{436}
$$

In other words, we need only combine linearly the I-values with the I-co-efficients. This formula plays a basic role in the theory of polynomial approximation, as we shall illustrate later. At this moment, we will content ourselves with the observation that the nth derivative of a polynomial of nth degree is a constant, and especially that it is equal to the product of $n!$ with the highest coefficient of the polynomial. Accordingly, we obtain from (436) the following formula of numerical differentiation

$$
P_n^{(n)} = n! \sum_{i=0}^{n} \lambda_i y_i. \tag{437}
$$

This equation permits calculation of the nth derivative of the I-polynomial from the $(n + 1)$ I-values y_i. In the case of *equidistant* I-places, it follows from (422) that

$$
P_n^{(n)} = \frac{1}{h^n} \sum_{\rho=0}^{n} \pm \binom{n}{\rho} y_\rho \tag{438}
$$

Here the signs alternate with increasing ρ and the last sign ($\rho = n$) is positive; for example:

$$
P_3^{(3)} = \frac{1}{h^3} \left(-y_0 + 3y_1 - 3y_2 + y_3 \right).
$$

The equation (438) for the nth derivative of any polynomial of nth degree, which has the values y_ρ, is precisely the rule for numerical differentiation which was deduced in 6.1 from a somewhat heuristic argument. This rule has accordingly found its final justification here.

In the equidistant case, the numerical differentiation will be facilitated by use of the following difference table. We shall introduce this *difference table* first by the numerical example of the somewhat expanded logarithm table (434)

x	y	Δ^1	Δ^2	Δ^3	Δ^4	Δ^5	
1.5	0.1761						
		1249					
2.0	0.3010		−280				
		969		103			
2.5	0.3979		−177		−48		(439)
		792		55		25	
3.0	0.4771		−122		−23		
		670		32			
3.5	0.5441		−90				
		580					
4.0	0.6021						

The first two columns contain the *I*-places and *I*-values. Every number in the column Δ^1 of *first differences* is formed by subtraction of the two adjacent values of the *y*-columns in the sequence "below minus above." (The first differences Δ^1 are given in units of the 4th place after the decimal point.) In precisely the same way, the column Δ^2 of the *second differences* consists of the differences of the column Δ^1, and so on. Using general and conventional notation this difference table looks as follows:

x	y	Δ^1	Δ^2	Δ^3	Δ^4	
x_0	y_0					
		$\Delta^1(0.5)$				
x_1	y_1		$\Delta^2(1)$			
		$\Delta^1(1.5)$		$\Delta^3(1.5)$		(440)
x_2	y_2		$\Delta^2(2)$		$\Delta^4(2)$	
		$\Delta^1(2.5)$		$\Delta^3(2.5)$		
x_3	y_3		$\Delta^2(3)$			
		$\Delta^1(3.5)$				
x_4	y_4					

and here then

$$\Delta^1(0.5) = y_1 - y_0, \quad \Delta^1(1.5) = y_2 - y_1, \quad \Delta^1(2.5) = y_3 - y_2, \quad \Delta^1(3.5) = y_4 - y_3,$$

$$\Delta^2(1) = y_0 - 2y_1 + y_2, \qquad \Delta^2(2) = y_1 - 2y_2 + y_3, \qquad \Delta^2(3) = y_2 - 2y_3 + y_4,$$

$$\Delta^3(1.5) = -y_0 + 3y_1 - 3y_2 + y_3, \qquad \Delta^3(2.5) = -y_1 + 3y_2 - 3y_3 + y_4,$$

$$\Delta^4(2) = y_0 - 4y_1 + 6y_2 - 4y_3 + y_4. \tag{441}$$

As in 6.1, we see how the binomial coefficients appear here again through the mechanism of the Pascal triangle. The difference table contains everything necessary for the numerical differentiation. Let y_i be the values in a table of the function $f(x)$, i.e.,

$$y_i = f(x_i).$$

According to the rules of numerical differentiation (6.1), we then find, for example,

$$f'(x_0 + 1.5h) \sim \frac{\Delta^1(1.5)}{h}, \quad f''(x_0 + 3h) \sim \frac{\Delta^2(3)}{h^2}, \quad f^{(4)}(x_0 + 2h) \sim \frac{\Delta^{(4)}(2)}{h^4},$$

where h is the table step. If the tabulated function is a polynomial of nth degree, then all numbers in the column of nth differences must be equal because the nth derivative of this polynomial is a constant.

But the difference table is also useful for many other purposes. It is often used in order to check the smoothness of a tabulated function. Errors and irregularities in the function values show themselves in the higher differences—which are of course essentially higher derivatives—much more markedly than in the function values themselves.

Now we shall derive additional classical *interpolation formulas*. For this purpose we first consider the general case of pairwise different but not necessarily equidistant I-places $x_0, x_1, x_2, \ldots x_n$. We shall proceed recursively and assume that we have already succeeded in constructing the interpolating polynomial $P_{n-1}(x)$ which passes through the I-values $y_0, y_1, \ldots, y_{n-1}$. The polynomial $P_n(x)$ is to be found which *additionally* also passes through the next I-value y_n. We assume this polynomial to be equal to

$$P_n(x) = P_{n-1}(x) + Q_n(x),$$

i.e., we add a polynomial of nth degree. Obviously, $Q_n(x)$ has to vanish at the points $x_0, x_1, \ldots, x_{n-1}$ and accordingly it must have the form

$$Q_n(x) = \alpha(x - x_0)(x - x_1) \ldots (x - x_{n-1})$$

where α is a constant as yet to be determined. Then

$$P_n(x) = P_{n-1}(x) + \alpha(x - x_0)(x - x_1) \ldots (x - x_{n-1}).$$

Differentiating n times we find that

$$P_n^{(n)} = n!\,\alpha, \qquad \text{i.e.,} \qquad \alpha = \frac{P_n^{(n)}}{n!},$$

and finally this gives the *general Newton interpolation formula*

$$P_n(x) = P_{n-1}(x) + \frac{(x - x_0)(x - x_1) \ldots (x - x_{n-1})}{n!}\,P_n^{(n)}. \qquad (442)$$

Here, according to equation (437), the number $P_n^{(n)}$ has to be calculated from the I-values $y_0, \ldots, y_n$ by numerical differentiation.

First application. Let the interpolating abscissas be equidistant and numbered consecutively as in the difference table (440). Linear interpolation in the interval (x_0, x_1) then yields first of all the polynomial of first degree

$$P_1(x) = y_0 + \frac{x - x_0}{h}\,\Delta^1(0.5)$$

which assumes the I-values y_0, y_1. For the polynomial of second degree $P_2(x)$—which additionally assumes y_2—we derive from (438) that

$$P_2'' = \frac{1}{h^2}(y_0 - 2y_1 + y_2) = \frac{\Delta^2(1)}{h^2}.$$

From (442) then follows the result

$$P_2(x) = y_0 + \frac{x - x_0}{h}\,\Delta^1(0.5) + \frac{(x - x_0)(x - x_1)}{2!\,h^2}\,\Delta^2(1).$$

And in exactly the same way we find in the next step the polynomial of 3rd degree

$$P_3(x) = y_0 + \frac{(x - x_0)}{h}\,\Delta^1(0.5) + \frac{(x - x_0)(x - x_1)}{2!\,h^2}\,\Delta^2(1)$$

$$+ \frac{(x - x_0)(x - x_1)(x - x_2)}{3!\,h^3}\,\Delta^3(1.5)$$

which assumes the interpolating values y_0, y_1, y_2, y_3. Continuing according to this evident rule, we obtain the *special Newton interpolation formula*. In table (440) it uses those differences located on the downward diagonal originating from the element y_0.

Second application. In the following discussion, we shall use a constant step h and I-places which are located on both sides of x_0. We therefore enlarge the difference scheme (440):

(440a)

x_0-2h	y_{-2}		$\Delta^2(-2)$		$\Delta^4(-2)$	
		$\Delta^1(-1.5)$		$\Delta^3(-1.5)$		$\Delta^5(-1.5)$
x_0-h	y_{-1}		$\Delta^2(-1)$		$\Delta^4(-1)$	
		$\Delta^1(-0.5)$		$\Delta^3(-0.5)$		$\Delta^5(-0.5)$
x_0	y_0		$\Delta^2(0)$		$\Delta^4(0)$	
		$\Delta^1(0.5)$		$\Delta^3(0.5)$		$\Delta^5(0.5)$
x_0+h	y_1		$\Delta^2(1)$		$\Delta^4(1)$	
		$\Delta^1(1.5)$		$\Delta^3(1.5)$		$\Delta^5(1.5)$
x_0+2h	y_2		$\Delta^2(2)$		$\Delta^4(2)$	

For the interpolation we shall now use these I-places in the following order:

$$x_0, \quad x_0+h, \quad x_0-h, \quad x_0+2h, \quad x_0-2h, \quad x_0+3h, \quad \dots$$

(alternating on the right and left of x_0). When the linear interpolation

$$P_1(x) = y_0 + \frac{x-x_0}{h}\,\Delta^1(0.5)$$

is extended to the quadratic interpolation $P_2(x)$, the added polynomial has to vanish at the points

$$x_0, (x_0+h).$$

Furthermore, the derivative P_2'' has to be formed by numerical differentiation at the points $x_0, (x_0+h), (x_0-h)$:

$$P_2'' = \frac{1}{h^2}\,[y(x_0-h) - 2y(x_0) + y(x_0+h)] = \frac{\Delta^2(0)}{h^2}.$$

Here the notations of the difference table (440) were used. Consequently, we now find that according to (442)

$$P_2(x) = y_0 + \frac{x-x_0}{h}\,\Delta^1(0.5) + \frac{(x-x_0)(x-x_0-h)}{2!\,h^2}\,\Delta^2(0).$$

Introducing

$$p = \frac{x-x_0}{h}$$

we can write this in the simpler form

$$P_2(x) = y_0 + p\Delta^1(0.5) + \frac{p(p-1)}{2!}\Delta^2(0).$$

When this is extended to $P_3(x)$, the additional polynomial has to vanish at the points x_0, $(x_0 + h)$, $(x_0 - h)$, i.e., for $p = 0, +1, -1$, and we have

$$P_3(x) = y_0 + p\Delta^1(0.5) + \frac{p(p-1)}{2!}\Delta^2(0) + \frac{p(p-1)(p+1)}{3!}\Delta^3(0.5).$$

(The 3rd difference is, of course, to be formed at the points x_0, $(x_0 + h)$, $(x_0 - h)$, $(x_0 + 2h)$ with the I-values now available.) Continuing in this way, we obtain the *first Gauss formula*. The *second Gauss formula* results from the grouping

$$x_0, \qquad x_0 + h, \qquad x_0 + 2h, \qquad x_0 - h, \qquad x_0 + 3h, \qquad x_0 - 2h,$$

and, if we proceed up to the third degree, it reads

$$P_3(x) = y_0 + p\Delta^1(0.5) + \frac{p(p-1)}{2!}\Delta^2(1) + \frac{p(p-1)(p-2)}{3!}\Delta^3(0.5).$$

If we terminate both Gauss formulas at the same difference of odd order, then they use the same I-places, and therefore represent the same I-polynomial, but only in different ways.

This polynomial can also be calculated by taking the arithmetic mean of the two Gauss formulas. This results in the *Bessel formula* which we shall work out to the 5th degree in order to bring out the rule governing its formation:

$$P_5(x) = y_0 + p\Delta^1(0.5) + \frac{p(p-1)}{2!}\frac{\Delta^2(0) + \Delta^2(1)}{2} + \frac{p(p-1)}{3!}\left(p - \frac{1}{2}\right)\Delta^3(0.5)$$

$$+ \frac{p(p-1)(p+1)(p-2)}{4!}\frac{\Delta^4(0) + \Delta^4(1)}{2}$$

$$+ \frac{p(p-1)(p+2)(p-2)}{5!}\left(p - \frac{1}{2}\right)\Delta^5(0.5).$$

In the practical application of interpolation, this formula is preferred over the previous one if the N-point x lies between x_0 and $x_1 = x_0 + h$. This formula uses only those differences which are located horizontally beside these two arguments and on the half row in between. If we introduce

$$q = 1 - p, \qquad r = p - \tfrac{1}{2}, \qquad L = y_0 + p\Delta^1(0.5),$$

and if we stop with the 3rd difference, we obtain

$$P_3(x) = L - \frac{pq}{2}\left[\frac{\Delta^2(0) + \Delta^2(1)}{2} + \frac{r}{3}\Delta^3(0.5)\right].$$

This convenient formula for cubic interpolation is equivalent to (432) and can be evaluated with the help of the nomogram in Appendix II, Table 8. When the r-term is neglected, an often used interpolation formula results (the so-called quadratic improvement of linear interpolation).

The *Everett interpolation formula* is a modification of the Bessel formula. It is obtained by expressing the differences of odd order in terms of those of even order:

$$\Delta^1(0.5) = y_1 - y_0, \qquad \Delta^3(0.5) = \Delta^2(1) - \Delta^2(0),$$

$$\Delta^5(0.5) = \Delta^4(1) - \Delta^4(0),$$

and so on. This results in

$$P_5(x) =$$

$$py_1 + \frac{(p+1)p(p-1)}{3!}\Delta^2(1) + \frac{(p+2)(p+1)p(p-1)(p-2)}{5!}\Delta^4(1)$$

$$-(p-1)y_0 - \frac{p(p-1)(p-2)}{3!}\Delta^2(0) - \frac{(p+1)p(p-1)(p-2)(p-3)}{5!}\Delta^4(0).$$

It is self-evident how the formula is to be continued if interpolation by higher degree polynomials is desired.

Third application. With a decreasing arrangement

$$x_0, \qquad x_0 - h, \qquad x_0 - 2h, \qquad x_0 - 3h, \ldots$$

of the I-places, we find as counterpart to the Newton formula the interpolation formula

$$P_n(x) = y_0 + p\Delta^1(-0.5) + \frac{p(p+1)}{2!}\Delta^2(-1)$$

$$+ \frac{p(p+1)(p+2)}{3!}\Delta^3(-1.5) + \ldots.$$

Here it is understood that we stop with the nth difference. Again we have

$$p = \frac{x - x_0}{h}.$$

This formula uses those differences which in table (440a) are located on the upward diagonal originating with y_0. Integration of this formula over the interval $(x_0 - h, x_0)$ yields

$$\int_{x_0-h}^{x_0} P_n(x)\, dx = h \int_{-1}^{0} P_n\, dp$$

$$= h\left\{y_0 - \frac{1}{2}\Delta^1(-0.5) - \frac{1}{12}\Delta^2(-1) - \frac{1}{24}\Delta^3(-1.5) - \frac{19}{720}\Delta^4(-2)\right.$$

$$\left. - \frac{3}{160}\Delta^5(-2.5) - \frac{863}{60480}\Delta^6(-3) - \frac{275}{24192}\Delta^7(-3.5) - \ldots\right\}.$$

If a function $f(x)$ is given at equidistant I-places, this formula can serve for the approximate integration of $f(x)$ over the "last" interval (see also Appendix II, Table 2).

By integrating our formula over the interval $(x_0, x_0 + h)$, we find in the same way

$$\int_{x_0}^{x_0+h} P_n(x)\, dx = h\left\{y_0 + \frac{1}{2}\Delta^1(-0.5) + \frac{5}{12}\Delta^2(-1) + \frac{3}{8}\Delta^3(-1.5) + \frac{251}{720}\Delta^4(-2)\right.$$

$$\left. + \frac{95}{288}\Delta^5(-2.5) + \frac{19087}{60480}\Delta^6(-3) + \frac{5257}{17280}\Delta^7(-3.5) + \ldots\right\}.$$

This is a formula for the integration over the "overhanging" interval (Appendix II, Table 2). With the Adams method it can be used for the solution of differential equations (compare 6.34).

7.2 Systems of Polynomials

Consider a sequence of polynomials with increasing degree:

$$P_0(x),\ P_1(x),\ \ldots,\ P_n(x),\ \ldots. \tag{443}$$

We call this a polynomial system if the polynomial $P_\rho(x)$ in this sequence has exactly the degree ρ, i.e., if the highest coefficient (the coefficient of x_ρ) does not vanish, and if this holds for all ρ. The main problem of this section will be to expand a given polynomial $Q_n(x)$ of nth degree in terms of the polynomials of this polynomial system, i.e., to represent it in the form

$$Q_n(x) = c_0 P_0(x) + c_1 P_1(x) + \ldots + c_n P_n(x). \tag{444}$$

With this we pursue the following aim: If we truncate the expansion, i.e., form the *partial sum*

$$c_0 P_0(x) + c_1 P_1(x) + \ldots + c_m P_m(x)$$

with $m < n$, then this partial sum as a whole is a polynomial of mth degree which might approximate the polynomial $Q_n(x)$ reasonably well. The success of this idea depends, of course, on the proper choice of the polynomial system $P_\rho(x)$.

In mathematical physics, polynomial systems are occasionally generated in a natural way if partial differential equations are solved by the method of separation of variables. We mention only the *Legendre polynomials* which occur when separation of variables is applied to potential-theoretic problems in spatial polar coordinates [28]. For our purposes of approximation the following polynomial system is especially suitable.

7.21   Chebyshev Polynomials

In the following the argument x shall always vary in the interval $(-1, +1)$ and possible I-places are also to be contained in this interval. If a different interval was given originally, it must be reduced to the above "normal interval" by means of translation and change in scale of the x-axis. Since the absolute value of x is now always less than or equal to 1, it is possible to introduce an angle φ by means of the relation

$$\cos \varphi = x. \tag{445}$$

To understand this geometrically, we can introduce the unit circle in the x, y-plane (i.e., the circle with the origin as center and with the radius 1). Then φ is the polar angle in this circle. We note that

$$\cos 2\varphi = \cos^2 \varphi - \sin^2 \varphi = -1 + 2x^2$$

is a quadratic polynomial in x. It is called the second *Chebyshev polynomial* (T-polynomial[1] for short), $T_2(x)$. We define the 0th and 1st T-polynomial simply by

$$T_0(x) = 1, \qquad T_1(x) = x. \tag{446}$$

We want to show generally that $\cos n\varphi$ ($n = 1, 2, 3, \ldots$) is a polynomial of nth degree in x, called the nth T-polynomial $T_n(x)$. The proof is obtained by induction using the trigonometric identity:

$$\cos (n + 1)\varphi + \cos (n - 1)\varphi = 2 \cos n\varphi \cos \varphi$$

i.e.,

$$T_{n+1}(x) + T_{n-1}(x) = 2x T_n(x)$$

or

$$T_{n+1}(x) = 2x T_n(x) - T_{n-1}(x). \tag{447}$$

---

[1] The T stands for Tschebyscheff—the German transliteration of the Russian mathematician's name (the translators).

Our statement follows quite easily from this recursion formula. In fact, if it has already been asserted that T_{n-1}, T_n are polynomials of degree $(n - 1)$, n, respectively, it then follows as a result of the multiplication with x that T_{n+1} is a polynomial of degree $(n + 1)$. At the same time, the recursion formula offers a convenient means of calculating the first T-polynomials:

$$T_2(x) = -1 + 2x^2, \qquad T_3(x) = -3x + 4x^3, \qquad T_4(x) = 1 - 8x^2 + 8x^4$$
$$\tag{448}$$
$$T_5(x) = 5x - 20x^3 + 16x^5, \qquad T_6(x) = -1 + 18x^2 - 48x^4 + 32x^6.$$

Furthermore, we note the following properties of T-polynomials, all of which follow from (447) without much need for further explanation. All coefficients

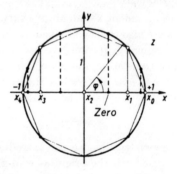

FIG. 35. T-abscissas x_k and zeros of $T_4(x)$.

of $T_n(x)$ are integers. The highest coefficient is 2^{n-1}. For even or odd n, $T_n(x)$ is an even or odd function of n, respectively. From the original definition

$$T_n(x) = \cos n\varphi \qquad \text{where} \qquad x = \cos \varphi \tag{449}$$

we further obtain the following facts:

$$T_n(1) = 1, \qquad T_n(-1) = (-1)^n. \tag{450}$$

In the interval $(-1, +1)$, the absolute value of $T_n(x)$ is ≤ 1. The *Chebyshev abscissas* (T-abscissas) are defined as those x-values for which this absolute value assumes its maximum 1. They are given by

$$|\cos n\varphi| = 1, \qquad \varphi = k\frac{\pi}{n}, \qquad k = 0, 1, \dots, n.$$

Hence, the T-abscissas are obtained by inscribing a regular $2n$-sided polygon in the unit circle, and then projecting its corners onto the x-axis (Fig. 35, $n = 4$). At these $(n + 1)$ T-abscissas, $T_n(x)$ therefore assumes alternately the values ± 1. At a point on the circle which halves the arc between two corners of the regular polygon we have $\cos n\varphi = 0$. The projection of such a point onto the x-axis is therefore a zero of $T_n(x)$. In this way we obtain a total of n zeros. It follows that all zeros of $T_n(x)$ are real and are contained in the interval $(-1, +1)$.

Appendix II, Table 9, provides a graphical representation of the first T-polynomials in the interval $(0, 1)$; compare also [29].

Instead of giving an explicit formula for the T-polynomials we rather want to develop the powers x^n in terms of the T-polynomials. The following identity holds for even integers $n = 2\nu$:

$$\cos^n \varphi = \frac{1}{2^{n-1}} \left[\cos n\varphi + \binom{n}{1} \cos (n-2)\varphi + \binom{n}{2} \cos (n-4)\varphi + \ldots \right.$$

$$\left. \ldots + \binom{n}{\nu-1} \cos 2\varphi \right] + \frac{1}{2^n} \binom{n}{\nu}.$$

This can be verified in the quickest way by using the binomial expansion of the expression

$$\cos^n \varphi = \left(\frac{e^{i\varphi} + e^{-i\varphi}}{2} \right)^n$$

From this formula we obtain

$$x^n = \frac{1}{2^{n-1}} \left[T_n + \binom{n}{1} T_{n-2} + \binom{n}{2} T_{n-4} + \ldots + \binom{n}{\nu-1} T_2 \right] + \frac{1}{2^n} \binom{n}{\nu} T_0$$

$$n \text{ even} = 2\nu. \tag{451}$$

Similarly, we find for odd n:

$$x^n = \frac{1}{2^{n-1}} \left[T_n + \binom{n}{1} T_{n-2} + \binom{n}{2} T_{n-4} + \ldots + \binom{n}{\nu} T_1 \right] \tag{452}$$

$$n \text{ odd} = 2\nu + 1.$$

We now come to a theorem about the range of values of polynomials. This theorem will form the basis for our theory of approximation. We consider a polynomial $P_n(x)$ of nth degree which first of all has the same highest coefficient 2^{n-1} as $T_n(x)$ and secondly, which satisfies the condition

$$|P_n(x)| < 1 \quad \text{for} \quad -1 \leq x \leq 1. \tag{453}$$

In other words, in the normal interval, $P_n(x)$ is completely contained inside the strip of the x, y-plane bounded by the two straight lines $y = \pm 1$, while $T_n(x)$ reaches the border of this strip in the $(n+1)$ T-abscissas. Now let x_i

be a T-abscissa for which $T_n(x)$ has the value $+1$. Considering (453) we then have the relations

$$T_n(x_i) = 1, \qquad P_n(x_i) < 1; \qquad \text{hence,} \qquad P_n(x_i) - T_n(x_i) < 0, \qquad (454)$$

and it is indeed geometrically clear that we have to proceed downward in order to come from the value of T_n to the value of P_n. Similarly, we find for a T-abscissa x_j in which T_n assumes its minimum (-1) that

$$T_n(x_j) = -1, \qquad P_n(x_j) > -1; \qquad \text{hence,} \qquad P_n(x_j) - T_n(x_j) > 0. \quad (455)$$

When we proceed through the T-abscissas from (-1) to $(+1)$, $T_n(x)$ will be alternately ± 1. It therefore follows that in these T-abscissas the difference function $P_n(x) - T_n(x)$ is alternately positive and negative. Hence, this function has at least n zeros in the interval $(-1, +1)$. But since P_n and T_n both have the same highest coefficient, this difference function is a polynomial of degree $(n - 1)$. As such, it can have n zeros only if it is identically zero. This means that we should have $P_n(x) \equiv T_n(x)$, but because of (453) this is impossible. In other words, a polynomial $P_n(x)$ of the desired type does not exist. The positive form of this negative statement naturally states: A polynomial of nth degree with highest coefficient 2^{n-1} must assume an absolute value ≥ 1 at least at one point in the interval $(-1, +1)$.

Now let $U_n(x)$ be any polynomial of nth degree with the highest coefficient $\gamma_n \neq 0$. Then the polynomial

$$U_n^*(x) = \frac{2^{n-1}}{\gamma_n} U_n(x)$$

has the highest coefficient 2^{n-1}. According to the result just proved, at least one point x then exists, such that

$$|U_n^*(x)| \geq 1; \qquad \text{hence,} \qquad |U_n(x)| \geq \frac{|\gamma_n|}{2^{n-1}}$$

In the interval $(-1, +1)$ a polynomial of nth degree with highest coefficient γ_n has to assume at least once an absolute value $\geq |\gamma_n|/2^{n-1}$. (The theorem also holds trivially for $\gamma_n = 0$.)

7.22 Approximation by Means of Chebyshev Expansion

Consider a polynomial of nth degree

$$Q_n(x) = a_0 + a_1 x + \ldots + a_n x^n. \qquad (456)$$

If every occurring power of x is expressed in terms of T-polynomials by means of formulas (451), (452), and if those expressions are combined, the result will be the expansion of $Q_n(x)$ in terms of T-polynomials

$$Q_n(x) = c_0 T_0(x) + c_1 T_1(x) + \ldots + c_n T_n(x). \qquad (457)$$

Here the c are combinations of the a. We further introduce the *partial sums* of this expansion:

$$S_m(x) = \sum_{k=0}^{m} c_k T_k(x), \qquad m = 0, 1, \ldots, n, \tag{458}$$

or, in particular,

$$S_n(x) = Q_n(x), \qquad S_{n-1}(x) = c_0 T_0(x) + c_1 T_1(x) + \ldots + c_{n-1} T_{n-1}(x) \tag{459}$$

and

$$Q_n(x) - S_{n-1}(x) = c_n T_n(x).$$

Hence, in the interval $(-1, +1)$

$$|Q_n(x) - S_{n-1}(x)| \leq |c_n| \tag{460}$$

holds. In other words, if $Q_n(x)$ is approximated by the expression resulting from cancellation of the last term in the expansion (457), the approximation error is at most equal to the absolute value of the last coefficient c_n. According to (457), $Q_n(x)$ has the highest coefficient $2^{n-1} c_n$. Now let $P_{n-1}(x)$ be any polynomial of $(n-1)$th degree. Then the difference $Q_n(x) - P_{n-1}(x)$ is a polynomial of nth degree with the same highest coefficient. According to our theorem on the range of values of polynomials, at least one point ξ must therefore exist in $(-1, +1)$, such that

$$|Q_n(\xi) - P_{n-1}(\xi)| \geq \frac{2^{n-1}|c_n|}{2^{n-1}} = |c_n|. \tag{461}$$

Hence, if P_{n-1} were used for the approximation of Q_n instead of S_{n-1}, the result would be no better. At least one point ξ would exist where the error of the approximation by P_{n-1} is at least as large as that of the approximation by S_{n-1}. This is seen by comparing (460) with (461). We say in brief that the partial sum $S_{n-1}(x)$ is the *best approximation* of the given polynomial $Q_n(x)$ by a polynomial of $(n-1)$th degree.

The polynomial $Q_n(x)$ is equal to the nth partial sum $S_n(x)$. Hence, S_{n-1} is the best approximation of S_n. Applying our considerations to S_{n-1} instead of S_n, we find that S_{n-2} is the best approximation of S_{n-1} by a polynomial of $(n-2)$th degree, and so on. *The T-expansion therefore has the fundamental property that every partial sum is the best approximation of the next partial sum.* We call such an expansion *optimal*.

However, we should certainly not conclude from this that the partial sum S_{n-2}, for example, is the best approximation of $Q_n(x)$ by a polynomial of $(n-2)$th degree. Here we add a brief consideration to this point. From

$$Q_n(x) = S_{n-2}(x) + c_{n-1} T_{n-1}(x) + c_n T_n(x)$$

we obtain the following error estimate for x in $(-1, +1)$:

$$|Q_n(x) - S_{n-2}(x)| \leq |c_{n-1}| \, |T_{n-1}(x)| + |c_n| \, |T_n(x)| \leq |c_{n-1}| + |c_n|. \tag{462}$$

It can be shown that for the approximation by any $P_{n-2}(x)$ there exists a ξ such, that

$$|Q_n(\xi) - P_{n-2}(\xi)| \geq |c_{n-1}|. \tag{463}$$

In case $|c_n|$ is small compared to $|c_{n-1}|$, there will be no big gap between the right sides of (462) and (463). While S_{n-2} is still not the best approximation of Q_n, it is certainly a very good one. This comment should make the following fact clear. Assume that the coefficients c_k decrease rapidly in the expansion (457), i.e., that c_{k+1} is small compared with c_k ($k = 0, 1, 2, \dots$). If the expansion is truncated with the term containing T_m, the partial sum $S_m(x)$ will be a good approximation of $Q_n(x)$ by a polynomial of mth degree. Such a decrease in the coefficients can be expected—at least from a certain point on—if the given polynomial $Q_n(x)$ has been obtained by truncation of a power series which has a radius of convergence larger than 1.

If we do want to have the very best approximation of $Q_n(x)$ by a polynomial of degree $m < n$, then the much more complicated methods of linear programming will have to be applied, (2.7).

Example. The function

$$f(x) = e^x$$

is to be approximated by polynomials in the interval $(-1, +1)$. We begin with the power series:

$$f(x) = 1 + x + 0.5x^2 + 0.16667x^3 + 0.04167x^4 + \dots . \tag{464}$$

The polynomial $Q_4(x)$ written here is now to be expressed in terms of T-polynomials. From (451) we find that

$$x^2 = 0.5T_0 + 0.5T_2, \qquad x^4 = 0.375T_0 + 0.5T_2 + 0.125T_4$$

and from (452) that

$$x^3 = 0.75T_1 + 0.25T_3.$$

Taken altogether this leads to

$$Q_4(x) = 1.2656T_0 + 1.1250T_1 + 0.2708T_2 + 0.0417T_3 + 0.0052T_4. \tag{465}$$

Toward the end the coefficients decrease very rapidly. The best approximation of Q_4 by a polynomial of 3rd degree comes about by dropping the last term. The maximal error is 0.0052, while dropping the last term in (464) leads to a maximal error of 0.04167, which is 8 times larger. A reasonable approximation of Q_4 by a polynomial of 2nd degree is given by the partial sum

$$S_2(x) = 1.2656T_0 + 1.1250T_1 + 0.2708T_2.$$

The maximal error is at most equal to

$$0.0417 + 0.0052 = 0.0469.$$

But the question of how well these polynomials approximate the originally given exponential function needs to be investigated separately.

There are many reasons for not giving a polynomial in numerical mathematics in its usual form of an expansion (456) in terms of powers of x, but in the form of its expansion (457) in terms of T-polynomials—and for always calculating with it in this latter form, even if no approximation problem is involved. The magnitude of the individual terms can then be seen at once, while the optimality property of the expansion often prevents a loss of decimal places. However, all this holds only for the normal interval $(-1, +1)$. Outside of this interval, the T-polynomials grow very rapidly and our approximation theorems are no longer valid.

In order to work this plan out consistently, it is necessary to build a "T-algebra" of polynomials where it is possible, for example, to multiply two T-expansions, and so on.

Here we only give the derivation of an algorithm for the calculation of the value of a T-expansion at a numerically given point x. (For the expansion (456) in terms of the powers of x this is accomplished by the *Horner* scheme, (4.51).) Let

$$Q_n(x) = c_0 T_0(x) + \ldots + c_{n-2} T_{n-2}(x) + c_{n-1} T_{n-1}(x) + c_n T_n(x). \qquad (466)$$

When we substitute for $T_n(x)$ the recursion formula (447)

$$T_n(x) + 2x T_{n-1}(x) - T_{n-2}(x) \qquad (467)$$

we find that

$$Q_n(x) = c_0 T_0 + \ldots + c_{n-3} T_{n-3} + (c_{n-2} - c_n) T_{n-2} + (c_{n-1} + 2x c_n) T_{n-1}. \qquad (468)$$

The coefficients up to and including c_{n-3} have not been modified. Since x is known numerically, we can again regard (468) as a T-expansion and in particular as one which is shorter by one term. With it we can repeat the process, and so on. At the end, it must be kept in mind that (467) is no longer correct for $n = 1$, and that instead

$$T_1(x) = x T_0(x). \qquad (469)$$

The calculation can best be worked out in the following modified version of the Horner scheme:

$$
\begin{array}{cccccc}
c_n & c_{n-1} & \boxed{c_{n-2}} & \boxed{c_{n-3}} & \ldots & \\
 & & -d_n & -d_{n-1} & & \\
 & 2x d_n & 2x d_{n-1} & & & (470) \\
\hline
d_n & \boxed{d_{n-1}} & d_{n-2} & & &
\end{array}
$$

The first row contains the coefficients of the given expansion. The highest coefficient is carried over unchanged into the 4th row and is denoted by d_n. Then we multiply by $2x$ and write the product on the right above d_n in the 3rd row. At the same time we enter $(-d_n)$ below c_{n-2} into the 2nd row. By addition of the second column d_{n-1} is then obtained. The coefficients of the shortened expansion (468) now are the boxed numbers and this shortened expansion can be treated in exactly the same way. The next step was still carried out in (470). Because of (469) we have to multiply at the end by x instead of by $2x$. The process will become completely clear if, as an example, we calculate the value of (465) for $x = 0.05$:

	52	417	2708	11250	12656
			-52	-422	-2698
		5	42	270	$\boxed{555}$

$2x = 0.1$	52	422	2698	11098	10513

At the end we proceeded here as follows:

$$555 = 0.05 \cdot 11098.$$

Hence we have

$$Q_4(0.05) = 1.0513,$$

and in comparison

$$e^{0.05} = 1.0513.$$

7.23 Expansion of the Interpolating Polynomial

As in 7.1, assume that the I-places $x_0, x_1, \ldots, x_n$ on the x-axis and the corresponding I-values $y_0, y_1, \ldots, y_n$ have been given. We now denote the I-polynomial by $Q_n(x)$. In addition, consider a system of polynomials $P_0(x), P_1(x), \ldots P_n(x)$. (For example, the successive powers of x, the Legendre or the T-polynomials). The I-polynomial is to be expanded in terms of this system; i.e., it is to be represented in the form

$$Q_n(x) = c_0 P_0(x) + c_1 P_1(x) + \ldots + c_{n-1} P_{n-1}(x) + c_n P_n(x) \quad (471)$$

For this we shall need the values $P_k(x_i)$ of the P-polynomials at the I-places. The tool for solving our problem will be the comparison of the highest coefficients on the right and left of equation (471). Hereby it should be noted that the highest coefficient on the right is provided by the last term $c_n P_n(x)$, because all previous terms contain at most the power x^{n-1}. $Q_n(x)$ has the I-values y_i; according to (436) its highest coefficient is therefore equal to

$$\sum_{i=0}^{n} \lambda_i y_i,$$

where the λ_i are the I-coefficients corresponding to the given I-places. $P_n(x)$ has the I-values $P_n(x_i)$; therefore, its highest coefficient is

$$\sum_{i=0}^{n} \lambda_i P_n(x_i).$$

From (471) then follows the equation

$$\sum \lambda_i y_i = c_n \sum \lambda_i P_n(x_i) \qquad \text{or} \qquad c_n = \frac{\sum \lambda_i y_i}{\sum \lambda_i P_n(x_i)}, \tag{472}$$

and with this we have found the last coefficient in the desired expansion. We see that the I-coefficients λ_i are not even needed, but only quantities proportional to them. In fact, a proportionality factor will cancel in the formula for c_n. Next, we introduce the partial sums

$$S_m(x) = \sum_{k=0}^{m} c_k P_k(x), \qquad m = 0, 1, \ldots, n. \tag{473}$$

In particular, we have

$$S_{n-1}(x) = Q_n(x) - c_n P_n(x).$$

This polynomial of $(n-1)$th degree has the I-values

$$S_{n-1}(x_i) = y_i - c_n P_n(x_i) \tag{474}$$

which can be calculated once c_n is known. Furthermore, S_{n-1} has $c_{n-1} P_{n-1}(x)$ as last term. We now treat S_{n-1} as we earlier did Q_n in order to determine c_{n-1}. In so doing we must not forget however that the equation (436) for the highest coefficient only holds if the number of I-points is exactly one larger than the degree of the corresponding interpolating polynomial. Since S_{n-1} has the degree $(n-1)$, only n I-points can be used for the evaluation of c_{n-1}. In other words, we have to drop one I-place; i.e., we have to go over to a reference system of the nth level, (7.1), and this reference system, of course, has different I-coefficients. Continuing in this way, we work out the expansion (471) from the right-hand side to the left until we have reached c_0.

Example. From a table for the exponential function we find

x	-1	-0.5	0	0.5	1
e^x	0.3679	0.6065	1.0000	1.6487	2.7183

The I-polynomial $Q_4(x)$ corresponding to these 5 I-values is to be expanded in terms of T-polynomials. We first copy from Appendix II, Table 3 the table of I-coefficients for 5 equidistant I-places and the corresponding reference systems. Then we add a table of the values of the first 4 T-polynomials at the given I-places (Appendix II, Table 9).

Levels	λ				
(5)	1	−4	6	−4	1
(4)	−1	2		−2	1
(3)	1		−2		1
(2)	−1				1

x	−1	−0.5	0	0.5	1	
T_4	1	−0.5	1	−0.5	1	12
T_3	−1	1	0	−1	1	6
T_2	1	−0.5	−1	−0.5	1	4
T_1	−1	−0.5	0	0.5	1	2

Outside on the right, we entered the respective sums

$$\sum \lambda_i T_k(x_i)$$

which will be needed in the denominators of the formulas (472). The calculation of the expansion coefficients c_k now proceeds as follows:

S	y					c
Q_4	0.3679	0.6065	1.0000	1.6487	2,7183	0.00545
	−545	272	−545	272	−545	
S_3	0.36245	0.60922	0.99455	1.65142	2.71285	0.04433
	4433	−4433		4433	−4433	
S_2	0.40678	0.56489	0.99455	1.69575	2.66852	0.27155
	−27155	13578	27155	13578	−27155	
S_1	0.13523	0.70067	1.26610	1.83153	2.39697	1.13087
	1.13087	56543		−56543	−1.13087	
S_0	1.26610	1.26610	1.26610	1.26610	1.26610	1.26610

This first row contains the given I-values. From these values together with the first row of the λ-table and using equation (472) we obtain the coefficient c_4 outside on the right. (The denominator in the equation equals 12.) The values of T_4, tabulated above, yield the corrections $-c_4 T_4(x_i)$ which were entered in the second row. According to (474), we then find by addition the values of the partial sum $S_3(x)$. From these, in turn, we can compute the coefficient c_3 with the help of the second row of the λ-table; hereby the center value of S_3 no longer plays a role. At the end a thorough control appears, since S_0 has to be precisely a constant. In this way we obtain the expansion

$$Q_4(x) = 1.2661\,T_0 + 1.1309\,T_1 + 0.2716\,T_2 + 0.0443\,T_3 + 0.0054\,T_4,$$

which deviates slightly from (465).

This method of expansion is of course better adapted to the problem of approximating the exponential function than is the method of telescoping the power series as was done in the last section. In fact, the power series approximates very well in the neighborhood of the origin, but very badly at the ends of the interval $(-1, +1)$.

If it is desirable to approximate more closely, the table step can be halved; i.e., 9 I-points can be used. The corresponding λ-table can be found in Appendix II, Table 3.

When this or some other method is applied to expand a polynomial $Q_n(x)$ in terms of *Legendre polynomials* $P_k(x)$, the mth partial sum $S_m(x)$ is an approximation of $Q_n(x)$ in the sense of the *method of least squares* (3.1). In fact, the integral

$$\int_{-1}^{+1} [Q_n(x) - S_m(x)]^2\,dx$$

is then smaller than any integral obtained by replacing $S_m(x)$ by an arbitrary polynomial of mth degree. This fact holds for

$$m = 0, 1, 2, \dots , (n - 1).$$

The proof is a result of the theory of systems of orthogonal polynomials [30].

7.3 Interpolation Problems in the Complex Plane

In this section we shall use I-places which are located in the complex z-plane. As in the real case, we shall try to place a polynomial in z through corresponding (also complex) I-values. However, we will consider only a special configuration of I-places, namely, that in which the I-places are the corners of a regular polygon.

A complex number ω is called an nth root of unity if it satisfies the relation

$$\omega^n = 1. \tag{475}$$

n is a positive integer. Let m be another (positive) integer. By raising the above equation to the power m, we see that, together with ω, ω^m is also an nth root of unity.   From

$$\omega^n - 1 = (\omega - 1)(1 + \omega + \omega^2 + \dots + \omega^{n-1}) = 0$$

it follows that a root of unity *different from* 1 satisfies the equation

$$1 + \omega + \omega^2 + \dots + \omega^{n-1} = 0 \qquad \text{or} \qquad \sum_{k=0}^{n-1} \omega^k = 0. \tag{476}$$

Considered as points in the complex plane, the successive powers of ω form a regular subdivision of the unit circle which closes itself as a result of (475), (possibly after several rotations around the circle).   Equation (476) states that the center of gravity of these partition points is located at the origin. Because ω^m is also an nth root of unity, it follows from (476) that

$$\sum_{k=0}^{n-1} \omega^{mk} = 0 \qquad \text{if} \qquad \omega^m \neq 1, \tag{477}$$

i.e., provided that ω does not accidentally happen to be an mth root of unity as well.   If ω does possess this property we trivially have

$$\sum_{k=0}^{n-1} \omega^{mk} = n \qquad \text{for} \qquad \omega^m = 1. \tag{477a}$$

In the following we shall use exclusively the nth root of unity:

$$\begin{aligned}
\omega &= e^{-2\pi i/n} = \cos\frac{2\pi}{n} - i\sin\frac{2\pi}{n} \\[2mm]
\omega^k &= e^{-2k\pi i/n} = \cos\frac{2k\pi}{n} - i\sin\frac{2k\pi}{n}.
\end{aligned} \tag{478}$$

The successive powers of ω form a regular polygon with its corners on the unit circle (Fig. 36, $n = 6$).   In this case ω can only be an mth root of unity if $(2m\pi/n)$ is a multiple of 2π, i.e., if m is a multiple of n.   Hence, if m is an integer ≥ 0, it follows from (477) and (477a) that

$$\sum_{k=0}^{n-1} \omega^{mk} = \begin{cases} n & \text{if } m = 0, n, 2n, \dots \\ 0 & \text{otherwise.} \end{cases} \tag{479}$$

For the case $n = 3$ we obtain in this way

$$m = 1: \quad 1 + \omega + \omega^2 = 0$$

$$m = 2: \quad 1 + \omega^2 + \omega^4 = 0 \quad \text{but for } m = 3: \quad 1 + \omega^3 + \omega^6 = 3. \tag{480}$$

$$m = 4: \quad 1 + \omega^4 + \omega^8 = 0$$

We now use the corners $1, \omega, \omega^2, \ldots, \omega^{n-1}$ of our regular polygon as I-places. Let y_k be given I-values and assume that they are associated with the I-places as follows:

$$
\begin{array}{lcccccc}
I\text{-places:} & 1 & \omega & \omega^2 & \cdots & \omega^{n-2} & \omega^{n-1} \\[4pt]
I\text{-values:} & y_n & y_{n-1} & y_{n-2} & \cdots & y_2 & y_1 \,.
\end{array}
\tag{481}
$$

If we go around the circle in the positive direction we encounter the I-values in the natural order $y_1, y_2, \ldots, y_n$. For the sake of uniformity we shall denote y_n also by y_0 (Fig. 36).

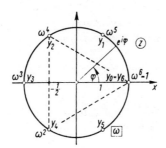

FIG. 36. Harmonic analysis.

Let z now be the complex variable in our Gauss plane. Then the interpolation problem of this section consists in constructing the polynomial of $(n-1)$st degree

$$
P_{n-1}(z) = c_0 + c_1 z + c_2 z^2 + \ldots + c_{n-1} z^{n-1}
\tag{482}
$$

which assumes the given I-values at the given I-places. We need only discuss this problem for the case $n = 3$, since the considerations proceed in exactly the same way for higher n. Thus we must find

$$
P_2(z) = c_0 + c_1 z + c_2 z^2,
\tag{483}
$$

such that the conditions

$$
\left.
\begin{array}{l}
y_3 = c_0 + c_1 + c_2 \\[4pt]
y_2 = c_0 + c_1 \omega + c_2 \omega^2 \\[4pt]
y_1 = c_0 + c_1 \omega^2 + c_2 \omega^4
\end{array}
\right|
\tag{484}
$$

are satisfied. We claim that the solution is simply given by

$$3c_0 = y_0 + y_1 \quad + y_2$$
$$3c_1 = y_0 + y_1\omega + y_2\omega^2$$
$$3c_2 = y_0 + y_1\omega^2 + y_2\omega^4$$

$$
\begin{array}{ccc}
1 & 1 & 1 \\
1 & \omega & \omega^2 \\
1 & \omega^2 & \omega^4
\end{array}
\qquad (485)
$$

To simplify verification of this statement, we have written the matrix of the coefficients on the right. The three equations (485) are then multiplied by the numbers in the first column of the matrix and the results are added. (Since the first column consists of "ones" only, this simply means addition.) Because of the relations (480) it follows that

$$3c_0 + 3c_1 + 3c_2 = 3y_0,$$

and in view of $y_0 = y_3$, this is identical with the first equation (484). Now we use the numbers in the second column of the matrix as multipliers and so obtain again because of (480):

$$3c_0 + 3c_1\omega + 3c_2\omega^2 = 3y_2$$

which is identical with the second equation (484). Similarly, we verify the third equation (484) by using the third column of the matrix.

Evaluation of the expansion coefficients by means of (485) is called the "*analysis*" *procedure*, while calculation of the values (484) of the polynomial is called "*synthesis*" *procedure*. It is surprising that both operations are performed with precisely the same coefficient matrix. Exactly the same follows from the relations (479) for arbitrary n. The rule of forming the matrix too is very simple. The second row contains the successive powers of ω and the third row those of ω^2. For higher n we have to enter into the fourth row the successive powers of ω^3, and so on. The numerical values of these matrix elements are obtained from (478).

For purposes of illustration, we here write out the formulas for $n = 6$. The corresponding root of unity is

$$\omega = \cos\frac{\pi}{3} - i\sin\frac{\pi}{3} = \frac{1}{2} - i\frac{\sqrt{3}}{2}. \qquad (486)$$

The interpolating polynomial has the general form

$$P_5(z) = c_0 + c_1 z + c_2 z^2 + c_3 z^3 + c_4 z^4 + c_5 c^5. \qquad (487)$$

Using the matrix notation of our Chapter 1, we have:

S		c_0	c_1	c_2	c_3	c_4	c_5	
	A	y_0	y_1	y_2	y_3	y_4	y_5	
$y_6 =$	$6c_0 =$	1	1	1	1	1	1	
$y_5 =$	$6c_1 =$	1	ω	ω^2	ω^3	ω^4	ω^5	(488)
$y_4 =$	$6c_2 =$	1	ω^2	ω^4	ω^6	ω^8	ω^{10}	
$y_3 =$	$6c_3 =$	1	ω^3	ω^6	ω^9	ω^{12}	ω^{15}	
$y_2 =$	$6c_4 =$	1	ω^4	ω^8	ω^{12}	ω^{16}	ω^{20}	
$y_1 =$	$6c_5 =$	1	ω^5	ω^{10}	ω^{15}	ω^{20}	ω^{25}	

The letter "A" signifies the notation for the analysis procedure and the letter "S" that for the synthesis procedure. The matrix is *symmetric*.

7.31 Reality Properties

The following considerations are based on the matrix (488); however, they can be carried out in a similar way for any *even n*. Two powers of ω, whose exponents complement each other to the value $n = 6$ (or to any multiple of 6), are located symmetrically to the x-axis in the complex plane (Fig. 36); in other words, they are *conjugate-complex*. If we now consider, for example, the rows c_2 and c_4 in (488), we find that corresponding elements in every column are conjugate-complex. The following therefore are pairs of conjugate-complex numbers:

$$\omega^2, \omega^4; \quad \omega^4, \omega^8; \quad \omega^6, \omega^{12}; \quad \omega^8, \omega^{16}; \quad \omega^{10}, \omega^{20}.$$

We now *assume* that the given I-values are *real*. From the formulas "A" it then immediately follows that the coefficients c_2, c_4 are conjugate-complex, because the respective contributions to c_2, c_4 on the part of the individual y-columns are conjugate-complex.

We shall denote the transition to the conjugate-complex value by an asterisk. Then

$$c_4 = c_2^* \quad \text{and similarly} \quad c_5 = c_1^*. \tag{489}$$

For any even n we evidently have the following rule: Two coefficients whose indices complement each other to n are conjugate-complex. For real I-values the calculation "A" can therefore be abbreviated as follows:

A	y_0	y_1	y_2	y_3	y_4	y_5
$6c_0 =$	1	1	1	1	1	1
$6c_1 =$	1	ω	ω^2	-1	ω^4	ω^5
$6c_2 =$	1	ω^2	ω^4	1	ω^2	ω^4
$6c_3 =$	1	-1	1	-1	1	-1

$$c_4 = c_2^* \qquad (490)$$
$$c_5 = c_1^*$$

Here we have furthermore made all exponents smaller than 6 by using the relation $\omega^6 = 1$. In addition, we note that $\omega^3 = -1$ (Fig. 36). Obviously, c_0, c_3 are real. An analogous simplification is possible for the synthesis procedure. For example, according to (488) we have

$$y_5 = c_0 = (\omega c_1 + \omega^5 c_1^*) + (\omega^2 c_2 + \omega^4 c_2^*) - c_3,$$

and the values in the brackets are real since they are the sums of conjugate-complex quantities; for instance, the first bracket contains double the real part of (ωc_1). For the synthesis procedure, we can even use the same matrix (490) again, but in transposed notation (compare 1.12):

S	$y_6 =$	$y_5 =$	$y_4 =$	$y_3 =$	$y_2 =$	$y_1 =$
c_0	1	1	1	1	1	1
$2c_1$	1	ω	ω^2	-1	ω^4	ω^5
$2c_2$	1	ω^2	ω^4	1	ω^2	ω^4
c_3	1	-1	1	-1	1	-1

$$(491)$$

We must simply remember that for the subsequent evaluation the complex numbers y_1, y_2, ... , y_6, calculated from this scheme, have to be replaced by their *real parts*.

Two remarks in conclusion: First, we can make use of the principle of *repeated halving* which has served us so well for the numerical integration (compare 6.22). For example, assume that as a first step the analysis procedure with 3 I-points is to be worked out by means of the formulas (485). In that case, we already know in (490) the partial sums

$$
\begin{aligned}
\text{for } 6c_0: \quad & y_0 + y_2 + y_4 \\
\text{for } 6c_1: \quad & y_0 + \omega^2 y_2 + \omega^4 y_4 \\
\text{for } 6c_2: \quad & y_0 + \omega^4 y_2 + \omega^2 y_4 \\
\text{for } 6c_3: \quad & y_0 + y_2 + y_4
\end{aligned}
$$

In other words, we know everything which results in (490) from the underlined columns. In order to see this, we need only note that the ω of (485) is the square of the ω of (490) and that the I-values in these two formulas are numbered differently. Figure 36 will make this completely clear; there the triangle inscribed in the hexagon corresponds to our present consideration. The analysis procedure with n I-points therefore halves the work for the analysis procedures with $2n$ I-points.

In the second place, we note the special case where the I-values show a symmetric behavior with respect to the x-axis of the complex plane. For the case $n = 6$ this means

$$y_1 = y_5, \qquad y_2 = y_4.$$

As a result, we can simplify (490); for example,

$$6c_1 = y_0 + (\omega + \omega^5)y_1 + (\omega^2 + \omega^4)y_2 - y_3.$$

Now (see Fig. 36),
$$\omega + \omega^5 = 1, \qquad \omega^2 + \omega^4 = -1.$$

Altogether, therefore, we obtain the matrix

	y_0	y_1	y_2	y_3
$6c_0 =$	1	2	2	1
$6c_1 =$	1	1	-1	-1
$6c_2 =$	1	-1	-1	1
$6c_3 =$	1	-2	2	-1

(492)

in which everything is real. Also for larger even n we obtain in this way real matrices and real coefficients c_k.

7.32 Harmonic Analysis (Fourier Analysis)

Exactly as in (481), consider a set of I-places and corresponding *real* I-values y_k. The number n of the I-places shall be even. The problem now is to find, instead of a polynomial, a rational function $R(z)$, which passes through these I-points. The form of $R(z)$ will be described first for the case $n = 6$:[1]

$$R(z) = c_0 + c_1 z + c_2 z^2 + \tfrac{1}{2} c_3 z^3$$
$$+ c_5 z^{-1} + c_4 z^{-2} + \tfrac{1}{2} c_3 z^{-3}. \tag{493}$$

The first row contains the positive, the second row the negative powers of z. For the general case $n = 2\nu$ we have analogously

$$R(z) = c_0 + c_1 z + c_2 z^2 + \ldots + c_{\nu-1} z^{\nu-1} + \tfrac{1}{2} c_\nu z^\nu$$
$$+ c_{n-1} z^{-1} + c_{n-2} z^{-2} + \ldots + c_{\nu+1} z^{-\nu+1} + \tfrac{1}{2} c_\nu z^{-\nu}. \tag{494}$$

However, we will stick to the example $n = 6$. By substituting the I-places $1, \omega, \ldots, \omega^5$ in (493), we find, because of $\omega^6 = 1$, the following equations for the coefficients c_k:

$$y_0 = y_6 = c_0 + c_1 + c_2 + c_3 + c_4 + c_5$$
$$y_5 = c_0 + c_1 \omega + c_2 \omega^2 + c_3 \omega^3 + c_4 \omega^4 + c_5 \omega^5$$
$$y_4 = c_0 + c_1 \omega^2 + c_2 \omega^4 + c_3 \omega^6 + c_4 \omega^8 + c_5 \omega^{10}.$$

We also find three additional equations which we will not bother to write down since it becomes obvious that we obtain exactly those equations given in (488) under the heading "S." It follows from this that the coefficients c_k for the rational interpolation problem (493) are exactly the same as those for the polynomial interpolation problem (487). They can therefore be calculated with the help of the reduced matrix (490). More especially, it follows from (489) that the coefficients which stand below each other in (493) are conjugate-complex.

Consider now, in particular, a point z on the unit circle, e.g.,

$$z = e^{i\varphi} \tag{495}$$

where φ is the polar angle in the unit circle. Then (493) can be written in the form

$$R(z) = c_0 + (c_1 z + c_1^* z^{-1}) + (c_2 z^2 + c_2^* z^{-2}) + \tfrac{1}{2} c_3 (z^3 + z^{-3}).$$

With

$$c_k = a_k + i b_k \tag{496}$$

[1] In complex function theory, the equivalent process is the transition from the *Taylor* series to the *Laurent* series.

we introduce the real and imaginary parts of our coefficients. Then

$$R(z) = a_0 + a_1(z + z^{-1}) + ib_1(z - z^{-1}) + a_2(z^2 + z^{-2}) + ib_2(z^2 - z^{-2})$$
$$+ \tfrac{1}{2}a_3(z^3 + z^{-3}).$$

(Note that c_0, c_3 are real.)   Because of (495) it now follows that

$$R(z) = a_0 + 2a_1 \cos \varphi - 2b_1 \sin \varphi + 2a_2 \cos 2\varphi - 2b_2 \sin 2\varphi + a_3 \cos 3\varphi.$$

In the general case, one obtains in the same way:

$$R(z) = a_0 + 2[a_1 \cos \varphi + a_2 \cos 2\varphi + \ldots + a_{\nu-1} \cos (\nu - 1)\varphi] + a_\nu \cos \nu\varphi$$
$$(497)$$
$$- 2[b_1 \sin \varphi + b_2 \sin 2\varphi + \ldots + b_{\nu-1} \cos (\nu - 1)\varphi].$$

Hence, $R(z)$ is a real function on the unit circle and in particular a *trigonometric polynomial* in φ. From now on, we will call it $\pi_n(\varphi)$, since it is a periodic function of φ with the period 2π.

This result can be formulated comprehensively in the language of real analysis as follows:

The interval $(0, 2\pi)$ of a real variable φ has been partitioned into n equal parts. In this way, the n I-places

$$\varphi = 0, \quad \frac{2\pi}{n}, \quad 2\frac{2\pi}{n}, \quad 3\frac{2\pi}{n}, \ldots, (n - 1)\frac{2\pi}{n}$$

appear. Let n be even and $= 2\nu$. In addition, let the corresponding I-values

$$y_0, \quad y_1, \quad y_2, \quad \ldots, \quad y_{n-1}$$

be given. Then the trigonometric polynomial $\pi_n(\varphi)$ represented by (497) assumes these I-values. The calculation of the coefficients is performed by using a matrix of the type (490) and by observing (496). The matrix is to be generated by following the rule formulated above—before formula (486)—and by using the root of unity

$$\omega = e^{-2\pi i/n}.$$

If the I-values are, in particular, the values of a periodic function with the period 2π, then $\pi_n(\varphi)$ approximates this function by a superposition of harmonic oscillations of the frequencies $0, 1, 2, \ldots, \nu$. There is, of course, no sense in trying to approximate non-periodic functions in this way.

Appendix II, Table 4, gives the computational directions for an analysis procedure with 12 I-points. The matrices in that table were generated by separating real and imaginary parts in the corresponding complex matrices of type (490). Furthermore, I-values located symmetrically to the x-axis were taken together. As an illustration of the halving method, Table 5 shows the extension of this analysis to 24 I-points. Tables 6 and 7 contain the corresponding computational directions for the synthesis procedure, i.e., for the evaluation of (497) for given coefficients a_k, b_k. The matrices are identical throughout with those of the analysis procedure, but they have to be read in transposed notation.

7.33   Application to the Expansion in Terms of Chebyshev Polynomials

As in the section on T-polynomials, (7.2), we are interested here in approximating and expanding real functions defined over the interval $(-1, +1)$ on the x-axis. As before, we introduce the unit circle with this interval as diameter. But we now consider the plane of this circle as the Gaussian plane of our complex variable z (Fig. 35). For $n = 2\nu$ our I-places (481),

$$1, \qquad \omega, \qquad \omega^2, \dots, \omega^{n-2}, \qquad \omega^{n-1}$$

located on the unit circle form a regular 2ν-sided polygon. The projections of the corners onto the x-axis are the $(\nu + 1)$ T-abscissas, i.e., the places where the polynomial $T_\nu(x)$ assumes its extrema. These T-abscissas shall now be used as I-places. Hence, we have no regular subdivision of the x-axis, but the I-places accumulate toward the ends of the normal interval $(-1, +1)$. If a point moves along the upper half circle in the positive direction, starting from $x = 1$, its projection moves toward the left on the x-axis. For this reason we shall number *the T-abscissas in order of decreasing x*:

$$1 = x_0 > x_1 > x_2 > \qquad \dots \qquad > x_\nu = -1$$

This notation contradicts earlier notations, but that cannot be avoided if we want to be in agreement with the harmonic analysis. Let the corresponding I-values

$$y_0, \qquad y_1, \qquad \dots \qquad , y_\nu$$

be given. The interpolation polynomial $P_\nu(x)$ is to be constructed. For this purpose, we transplant the value y_k for $k = 0, 1, \dots, \nu$ to the two points of the unit circle which are precisely above and below the abscissa x_k. In this way, we obtain exactly our interpolation problem in the complex plane, and even the special case discussed in the paragraph preceding (492), namely the special case where the I-values show symmetry relative to the x-axis. We saw that in this case the coefficients c_k are real. The final formula (497) therefore reduces to

$$R(z) = c_0 + 2[c_1 \cos \varphi + c_2 \cos 2\varphi + \dots + c_{\nu-1} \cos (\nu - 1)\varphi] + c_\nu \cos \nu\varphi.$$

Because of the fundamental definition (449) of the T-polynomials this can also be written in the form

$$R(z) = c_0 + 2[c_1 T_1(x) + c_2 T_2(x) + \ldots c_{\nu-1} T_{\nu-1}(x)] + c_\nu T_\nu(x). \qquad (498)$$

Hence, $R(z)$ turns out to be a real polynomial of νth degree in $x = \cos \varphi$. Substituting one of the T-abscissas x_k for x, we find that the corresponding angle φ belongs to that corner of the regular polygon which is above x_k. But, according to the construction, $R(z)$ has the value y_k. The polynomial (498) therefore assumes the prescribed I-values and hence is the desired I-polynomial

$$P_\nu(z) = c_0 + 2[c_1 T_1(x) + c_2 T_2(x) + \ldots + c_{\nu-1} T_{\nu-1}(x)] + c_\nu T_\nu(x). \qquad (499)$$

Moreover, this polynomial is in the form of its T-expansion. Our method for the solution of approximation problems consisted in truncating such expansions prematurely. We therefore need not even calculate all co-efficients c_k, and, as a result, the use of the T-abscissas as I-places greatly simplifies all interpolation and approximation problems. Furthermore, it has the decided advantage that, in contrast to the use of equidistant I-places, the I-polynomial is not subject to any "flutter." This polynomial is safely fixed numerically all the way up to the ends of the interval $(-1, +1)$.

The example of $\nu = 3$ $(n = 6)$ shall conclude these discussions. According to (492) we then have:

$$c_0 = \tfrac{1}{6}(y_0 + 2y_1 + 2y_2 + y_3), \qquad c_1 = \tfrac{1}{6}(y_0 + y_1 - y_2 - y_3)$$

$$c_2 = \tfrac{1}{6}(y_0 - y_1 - y_2 + y_3), \qquad c_3 = \tfrac{1}{6}(y_0 - 2y_1 + 2y_2 - y_3),$$

and according to (499) we find for the I-polynomial ,

$$P_3(x) = c_0 + 2c_1 T_1(x) + 2c_2 T_2(x) + c_3 T_3(x).$$

The T-abscissas are

$$x_0 = 1, \qquad x_1 = \tfrac{1}{2}, \qquad x_2 = -\tfrac{1}{2}, \qquad x_3 = -1.$$

For $x = 0$ the polynomial assumes the value

$$P_3(0) = c_0 - 2c_2 = \tfrac{1}{6}(-y_0 + 4y_1 + 4y_2 - y_3)$$

which was found in another way in the Example 1 for the Lagrange interpolation, (7.11).

Appendix I. Computational Examples

Example I. *Inversion of a 4 × 4 matrix* (for the theory see 1.2).

	x_1	x_2	x_3	x_4	σ
$y_1 =$	6.438	−3.747	2.185	1.882	−5.758
$y_2 =$	2.136	1.522	5.210	−1.123	−6.745
$y_3 =$	1.867	1.246	−1.110	8.331	−9.334
$y_4 =$	−3.736	7.642	1.500	1.232	−5.638
	−0.2241	−0.1496	0.1332		1.1204

	x_1	x_2	x_3	y_3	σ	
$y_1 =$	6.0162	−4.0285	2.4357	0.2259	−3.6494	−1
$y_2 =$	2.3877	1.6900	5.0604	−0.1348	−8.0032	1
$x_4 =$	−0.2241	−0.1496	0.1332	0.1200	1.1204	−1
$y_4 =$	−4.0121	7.4577	1.6641	0.1479	−4.2577	−1
	0.5380		−0.2231	−0.0198	0.5709	

	x_1	y_4	x_3	y_3	σ	
$y_1 =$	3.8489	−0.5402	3.3345	0.3057	−5.9493	−4
$y_2 =$	3.2969	0.2266	4.6834	−0.1683	−7.0384	2
$x_4 =$	−0.3046	0.0201	0.1666	0.1230	1.0350	−1
$x_2 =$	0.5380	0.1341	−0.2231	−0.0198	0.5709	1
	−0.7040	−0.0484		0.0359	1.5028	

	x_1	y_4	y_2	y_3	σ	
$y_1 =$	1.5014	−0.7016	0.7120	0.4254	−0.9382	−10
$x_3 =$	−0.7040	−0.0484	0.2135	0.0359	1.5028	−2
$x_4 =$	−0.4219	−0.0282	0.0356	0.1290	1.2854	−1
$x_2 =$	0.6951	0.1449	−0.0476	−0.0278	0.2356	+2
	0.4673	−0.4742	−0.2833	0.6294		

	y_1	y_4	y_2	y_3	σ	
$x_1 =$	0.6660	0.4673	−0.4742	−0.2833	0.6249	7
$x_3 =$	−0.4689	−0.3774	0.5473	0.2353	1.0629	−8
$x_4 =$	−0.2810	−0.2254	0.2357	0.2485	1.0218	−4
$x_2 =$	0.4630	0.4697	−0.3772	−0.2247	0.6700	8

Remarks

1. Before beginning the EX-algorithm, one should make certain that the matrix elements have about the same order of magnitude. This can be accomplished by appropriate multiplication of the variables x and y by suitable powers of ten.

2. In order to guard against round-off errors we carry in the computation one or two more decimal places (after the decimal point) than were given in the initial data.

3. Outside on the right, the deviations of the row-sums from the required value 1 have been noted down (in units of the last decimal place).

Example 2. *Four linear equations. Gauss algorithm* (Theory 1.31).

	x_1	x_2	x_3	x_4	1	σ
	6.438	−3.747	2.185	1.882	−4.635	−2.123
	2.136	1.522	5.210	−1.123	−5.213	−2.532
	1.867	1.246	−1.110	8.331	−4.132	−6.202
	−3.736	7.642	1.500	1.232	−5.866	−0.772
$x_4 =$	−0.2241	−0.1496	0.1332		0.4960	0.7444

x_1	x_2	x_3	1	σ	
6.0162	−4.0285	2.4357	−3.7015	−0.7220	−1
2.3877	1.6900	5.0604	−5.7700	−3.3680	1
−4.0121	7.4577	1.6641	−5.2549	0.1451	−1

$x_2 =$ 0.5380   −0.2231   0.7046   −0.0195

x_1	x_3	1	σ	
3.8489	3.3345	−6.5400	−0.6434	0
3.2969	4.6834	−4.5792	−3.4010	1

$x_3 =$ −0.7040   0.9778   0.7262

x_1	1	σ	
1.5014	−3.2795	1.7781	0

2.1843   2.0047   −0.5599   −0.3680

Remark. The matrix is the same as in Example 1. The computed values of the unknowns are listed in the last row.

Example 3. *Four equations, compact Gauss algorithm* (Theory 1.34).

x_1	x_2	x_3	x_4	1	σ
4	−1	−1	−1	−9.87	8.87
−1	4	−1	−1	−2.56	1.56
−1	−1	4	−1	−0.42	−0.58
−1	−1	−1	4	5.36	−6.36

	x_2	x_3	x_4	1	σ
4	0.250	0.250	0.250	2.468	−2.218
−1	3.750	0.333	0.333	1.341	−1.007
−1	−1.250	3.333	0.500	1.369	−0.869
−1	−1.250	−1.666	2.501	0.426	0.574
3.472	2.010	1.582	0.426		

Remarks

1. The resulting values for the unknowns are in the last row.

2. As always when solving linear equations, we have chosen the coefficient of the control variable in the above table in such a way that the row sum is equal to zero.   Control in the lower table is constituted by the fact that in each row the sum of the numbers on the *right of the staircase line* must be $= 1$.

Example 4.   *Method of Bernoulli* (Theory 4.61).

Cubic equation $\qquad 2x^3 - 11x^2 + 13x - 4 = 0$

or $\quad -0.5x^3 + 2.75x^2 - 3.25x + 1 = 0$

	γ	q	e	q'
	1			
		3.25		
	3.25		−0.8462	
		2.4038		0.6642
	7.8125		−0.2338	
		2.1700		0.8483
Paper Strip	16.953		−0.0914	
		2.0786		0.9279
0.5	35.238		−0.0408	
		2.0378		0.9640
−2.75	71.809		−0.0193	
		2.0185		0.9726
3.25	144.95		−0.0093	
		2.0092		
	291.23			

$\omega_1 = 0.498 \qquad\qquad\qquad\qquad \omega_2 = 1.03$

Convergence acceleration:

$$\frac{1}{\omega_1} = 2.0185\left(1 + \frac{-0.0093}{2.0185 - 0.9726}\right) = 2.0005$$

$$\omega_1 = 0.4999.$$

Remark. The paper strip is drawn in that position for which the γ-value 291.23 is being calculated.

Example 5. *Method of Bernoulli, cubic equation* $2x^3 - 4x^2 + 3x - 1 = 0$.

γ	q	e	q'	$q_\rho + q'_\rho$	$q_{\rho-1} \cdot q'_\rho$
1					
	3				
3		−1.3333			
	1.6667		0.8334	2.5001	2.5002
5		−0.6667			
	1		1.2	2.2	2.0000
5		−0.8			
	0.2		1.8	2.0	1.8
1		−7.2			
	−7		8.8889	1.8889	1.7778
−7		9.1429			
	2.1429		−0.2679	1.8750	1.8753
−15		−1.1429			
	1		0.9333	1.9333	2.0000
−15		−1.0667			
	−0.0667				
1					

$$2.0000x^2 - 1.9333x + 1 = 0$$

$$\omega_1,\ \omega_2 = 0.483 \pm 0.516i$$

Example 6. *Rhombus algorithm* (Theory 4.64). Cubic equation as in Example 4:

$$2x^3 - 11x^2 + 13x - 4 = 0 \qquad \text{or} \qquad -0.5x^3 + 2.75x^2 - 3.25x + 1 = 0.$$

q	e	q'	e'	q''
<u>3.25</u>		<u>0</u>		<u>0</u>
	−0.84615		−0.18182	
2.40385		0.66433		0.18182
	−0.23384		−0.04976	
2.17001		0.84841		0.23158
	−0.09142		−0.01358	
2.07859		0.92625		0.24516
	−0.04074		−0.00359	
2.03785		0.96340		0.24875
	−0.01926		−0.00093	
2.01859		0.98173		0.24968
	−0.00937		−0.00024	
2.00922		0.99086		0.24992

$$\omega_1 = 0.4977 \qquad\qquad \omega_2 = 1.009 \qquad\qquad \omega_3 = 4.001$$

Remark. The underlined numbers in the e and e' columns have been obtained as follows:

$$\frac{2.75}{-3.25} = -0.84615, \qquad \frac{-0.5}{2.75} = -0.18182.$$

Example 7. *Computation of a characteristic polynomial, an eigenvalue and a corresponding eigenvector* (Theory 5.21).

Given equations

x_1	x_2	x_3	x_4	*Multipliers*		
2.846	−0.294	−0.157	0.106	0.856		
−0.377	2.858	−0.125	−0.041	0.152	−0.4353	−0.3808
−0.787	−0.819	0.754	0.257	0.288	−0.6602	
0.856	0.152	0.288	0.650			

Generation of the polynomial

x_1	x_2	x_3	x_4				
				1	λ	λ^2	λ^3	λ^4

x_1	x_2	x_3	x_4	1	λ	λ^2	λ^3	λ^4
0.856	0.152	0.288	0.650	−1				
2.1522	−0.0531	0.0638	0.1585	0.650	−1			
0.6837	−0.7034	−0.4434	−0.1518	−1.4757	3.1642	−1		
0.1436	−1.0883	0.0476	0.0156	0.8343	−3.9045	4.1842	−1	
$x_1 =$ −0.1776	−0.3364	−0.7593	1.1682					
−0.4353	−0.6602	−1.4757	3.1642	−1				
−0.8248	−0.6734	−0.6709	−0.6770	3.1642	−1			
−1.1138	−0.0007	−0.0934	1.0021	−3.9045	4.1842	−1		
$x_3 =$ −0.6593		−2.2352	4.7928	−1.5147				
−0.3808		0.8343	−3.9045	4.1842	−1			
−1.1133		−0.0918	0.9987	−3.9034	4.1842	−1		
$x_2 =$		2.1909	−10.2534	10.9879	−2.6261			
		−2.5309	12.4138	−16.1362	7.1078	−1		

characteristic polynomial

One zero is $\lambda = 0.9412$

A corresponding eigenvector equals:

| 0.0298 | 0.0846 | 0.8782 | 1.0000 |

Remarks. The identifications x_1, x_2, x_3, x_4 above the columns apply to the entire calculation down to the eigenvector.

It should be remembered that $(-\lambda)$ has to be added to each term in the diagonal of the given system of equations.

Example 8. *Computation of* $\pi = \int_0^1 \dfrac{4dx}{1 + x^2}$ *by means of numerical integration* (Theory 6.22).

Function table

f				M	T
4	2			3	3
3.2				3.2	3.1
3.76471	2.56			3.16236	3.13118
3.93846	3.50685	2.87640	2.26549	3.14680	3.13899
3.98444	3.86415	3.64413	3.35738		
3.03858	2.71618	2.40941	2.12890	3.14290	3.14094

Integration table

T	S	C	D	E
3				
3.1	3.13333			
3.13118	3.14157	3.14212		
3.13899	3.14159	3.14159	3.14158	
3.14094	3.14159	3.14159	3.14159	3.14159

$$\pi = 3.14159 \ldots$$

Remark. In order to save space, the last set of values in the function table has been split into two rows.

Example 9. *Computation of the Bessel function* $y = J_0(x)$ *and its derivative* $y' = -J_1(x)$ *by solving the differential equation*

$$y'' = -\frac{1}{x}y' - y, \qquad y(0) = 1, \qquad y'(0) = 0$$

using the second method of Runge-Kutta. Theory 6.42, particularly equations (310) and (316) through (319).

$$z = y', \qquad f(x, y, z) = z, \qquad g(x, y, z) = -\frac{z}{x} - y$$

	x	y	z	k	l	K, L
	0	1	0	0	−0.1	
	0.1	1	−0.05	−0.01	−0.1	−0.00997
$h = 0.2$	0.1	0.995	−0.05	−0.01	−0.099	−0.09950
	0.2	0.99	−0.099	−0.0198	−0.099	
	0.2	0.99003	−0.0995	−0.0199	−0.09851	
	0.3	0.98008	−0.14875	−0.02975	−0.09685	−0.02962
$h = 0.2$	0.3	0.97515	−0.14792	−0.02958	−0.09642	−0.09653
	0.4	0.96045	−0.19592	−0.03918	−0.09413	
	0.4	0.96041	−0.19603	−0.07841	−0.18813	
	0.6	0.92120	−0.29010	−0.11604	−0.17508	−0.11409
$h = 0.4$	0.6	0.90239	−0.28357	−0.11343	−0.17191	−0.17282
	0.8	0.84698	−0.36794	−0.14718	−0.15482	
	0.8	0.84632	−0.36885			

$x =$	0	0.2	0.4	0.8
$J_0 =$	1	0.9900	0.9604	0.8463
$J_1 =$	0	0.0995	0.1960	0.3688

Remark. One can try to enlarge the step length within a certain distance of the singular point $x = 0$. For this purpose we chose the double step length for the third step.

Appendix II.   Tables

Table I.   Numerical Differentiation, Equidistant *I*-Places

In order to compute approximately the derivative of a function $f(x)$ at the specially marked point, the *I*-values have to be multiplied with the numbers in the table and the results added.   The final result must then be divided by the number written down on the right side of the table (h is the distance of two neighboring *I*-places).

Example.   (*First formula below*):

$$f'(0) \sim \frac{1}{2h}[-3f(0) + 4f(h) - f(2h)]$$

Differentiation at the beginning of the I-interval

f'	-3	4	-1			$2h$
f'	-11	18	-9	2		$6h$
f''	2	-5	4	-1		h^2
f'	-25	48	-36	16	-3	$12h$
f''	35	-104	114	-56	11	$12h^2$
f'''	-5	18	-24	14	-3	$2h^3$

Differentiation in the middle of the I-interval:

	∘——◉——∘					
f'	-1	1				h
	∘———◉———∘					
f'	-1	0	1			$2h$
f''	1	-2	1			h^2
	∘——∘—◉—∘——∘					
f'	1	-27	$+27$	-1		$24h$
f''	1	-1	-1	1		$2h^2$
f'''	-1	3	-3	1		h^3
	∘———∘——◉——∘———∘					
f'	1	-8	0	8	-1	$12h$
f''	-1	16	-30	16	-1	$12h^2$
f'''	-1	2	0	-2	1	$2h^3$
$f^{(4)}$	1	4	6	-4	1	h^4

Table 2. Numerical Integration, Equidistant *I*-Places

In order to calculate approximately the integral of a function $f(x)$ over the interval marked by a double line, the I-values have to be multiplied by the integration weights in the table and the results then have to be added. Following this, the result must be divided by the number on the outside right and multiplied by the length of the integration interval.

Cotes' Formulas

Trapezoidal rule o════o 1 1	2
Simpson's rule o════o════o 1 4 1	6
o════o════o════o 1 3 3 1	8
o════o════o════o════o 7 32 12 32 7	90
o════o════o════o════o════o 19 75 50 50 75 19	288
Periodic function o════o════o════o════o════o 1 1 1 1 1 |0| ←————— Period —————→	5

Integration over the Last Interval

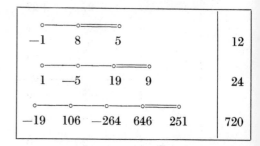

−1	8	5			12
1	−5	19	9		24
−19	106	−264	646	251	720

Overhanging Interval

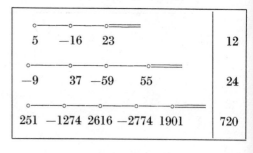

5	−16	23			12
−9	37	−59	55		24
251	−1274	2616	−2774	1901	720

Example. *(Simpson's formula).*

$$\int_{0}^{a} f(x)\, dx \sim \frac{a}{6}\left[f(0) + 4f\left(\frac{a}{2}\right) + f(a) \right].$$

Table 3. Interpolating Coefficients

5 equidistant I-places (step length h) and corresponding reference systems.

$$x_k = x_0 + kh$$

Level	x_0	x_1	x_2	x_3	x_4	Denominator
(5)	1	−4	6	−4	1	$24h^4$
(4)	−1	2		−2	1	$12h^3$
(3)	1		−2		1	$8h^2$
(2)	−1				1	$4h$

Example. *Reference system of the 4th level*

I-places: $\qquad\qquad x_0 \qquad x_1 \qquad x_3 \qquad x_4$

I-coefficients: $\qquad \dfrac{-1}{12h^3} \quad \dfrac{2}{12h^3} \quad \dfrac{-2}{12h^3} \quad \dfrac{1}{12h^3}$

9 equidistant I-places (step length h) and corresponding reference systems.

Level	x_0	x_1	x_2	x_3	x_4	x_5	x_6	x_7	x_8	Denominator
(9)	1	−8	28	−56	70	−56	28	−8	1	$40320h^8$
(8)	−1	6	−14	14		−14	14	−6	1	$10080h^7$
(7)	15	−64	84		−70		84	−64	15	$40320h^6$
(6)	−2	5		−7		7		−5	2	$1680h^5$
(5)	9	−16			14			−16	9	$2016h^4$
(4)	−1		2				−2		1	$96h^3$
(3)	1				−2				1	$32h^2$
(2)	−1								1	$8h$

For most purposes the denominator is not needed (Theory 7.1, 7.2).

Table 4. Harmonic Analysis, 12 I-Values

1. *Generation of sums and differences of symmetrically located I-values.*

| $y(0°)$ | $y(30°)$ | $y(60°)$ | $y(90°)$ | $y(120°)$ | $y(150°)$ | |
	$y(330°)$	$y(300°)$	$y(270°)$	$y(240°)$	$y(210°)$	$y(180°)$
s_0	s_1	s_2	s_3	s_4	s_5	s_6
	d_1	d_2	d_3	d_4	d_5	

For example, $s_1 = y(30°) + y(330°)$; $d_1 = y(30°) - y(330°)$.

2. *Matrix for the generation of the cosine terms.*

	s_0	s_1	s_2	s_3	s_4	s_5	s_6
$12a_0 =$	1	1	1	1	1	1	1
$12a_1 =$	1	u	0.5		-0.5	$-u$	-1
$12a_2 =$	1	0.5	-0.5	-1	-0.5	0.5	1
$12a_3 =$	1		-1		1		-1
$12a_4 =$	1	-0.5	-0.5	1	-0.5	-0.5	1
$12a_5 =$	1	$-u$	0.5		-0.5	u	-1
$12a_6 =$	1	-1	1	-1	1	-1	1

$$u = \cos 30° = 0.8660 \ldots$$

For the evaluation it is convenient to combine two symmetrically located quantities s_k, for example, s_0, s_6 or s_1, s_5. The same holds true for all following analysis problems.

3. *Matrix for the generation of the sine terms.*

	d_1	d_2	d_3	d_4	d_5
$12b_1 =$	-0.5	$-u$	-1	$-u$	-0.5
$12b_2 =$	$-u$	$-u$		u	u
$12b_3 =$	-1		1		-1
$12b_4 =$	$-u$	u		$-u$	u
$12b_5 =$	-0.5	u	-1	u	-0.5

4. *Trigonometric polynomial.*

$$\pi_{12}(\varphi) = a_0 + 2(a_1 \cos \varphi + \ldots + a_5 \cos 5\varphi)$$
$$+ a_6 \cos 6\varphi - 2(b_1 \sin \varphi + \ldots + b_5 \sin 5\varphi)$$

Table 5. Extension of an Analysis Problem with 12 *I*-Values to an Analysis Problem with 24 *I*-Values

1. *Sums and differences.*

$y(15°)$	$y(45°)$	$y(75°)$	$y(105°)$	$y(135°)$	$y(165°)$
$y(345°)$	$y(315°)$	$y(285°)$	$y(255°)$	$y(225°)$	$y(195°)$
σ_1	σ_2	σ_3	σ_4	σ_5	σ_6
δ_1	δ_2	δ_3	δ_4	δ_5	δ_6

For example, $\sigma_1 = y(15°) + y(345°)$; $\delta_1 = y(15°) - y(345°)$.

2. *Matrix for the generation of the cosine terms.*

	σ_1	σ_2	σ_3	σ_4	σ_5	σ_6		
$12\alpha_0 =$	1	1	1	1	1	1	$2a_0' = a_0 + \alpha_0,$	$2a_{12}' = a_0 - \alpha_0$
$12\alpha_1 =$	p	r	q	$-q$	$-r$	$-p$	$2a_1' = a_1 + \alpha_1,$	$2a_{12}' = a_1 - \alpha_1$
$12\alpha_2 =$	u		$-u$	$-u$		u	$2a_2' = a_2 + \alpha_2,$	$2a_{10}' = a_2 - \alpha_2$
$12\alpha_3 =$	r	$-r$	$-r$	r	r	$-r$	$2a_3' = a_3 + \alpha_3,$	$2a_9' = a_3 - \alpha_3$
$12\alpha_4 =$	0.5	-1	0.5	0.5	-1	0.5	$2a_4' = a_4 + \alpha_4,$	$2a_8' = a_4 - \alpha_4$
$12\alpha_5 =$	q	$-r$	p	$-p$	r	$-q$	$2a_5' = a_5 + \alpha_5,$	$2a_7' = a_5 - \alpha_5$

$$2a_6' = a_6$$

$$p = \cos 15° = 0.9659, \qquad q = \sin 15° = 0.2588$$
$$r = \sin 45° = 0.7071, \qquad u = \cos 30° = 0.8660$$

a_k from the analysis problem with 12 points

3. *Matrix for the generation of the sine terms.*

	δ_1	δ_2	δ_3	δ_4	δ_5	δ_6	
$12\beta_1 =$	$-q$	$-r$	$-p$	$-p$	$-r$	$-q$	$2b'_1 = \beta_1 + b_1, \quad 2b'_{11} = \beta_1 - b_1$
$12\beta_2 =$	-0.5	-1	-0.5	0.5	1	0.5	$2b'_2 = \beta_2 + b_2, \quad 2b'_{10} = \beta_2 - b_2$
$12\beta_3 =$	$-r$	$-r$	r	r	$-r$	$-r$	$2b'_3 = \beta_3 + b_3, \quad 2b'_9 = \beta_3 - b_3$
$12\beta_4 =$	$-u$		u	$-u$		u	$2b'_4 = \beta_4 + b_4, \quad 2b'_8 = \beta_4 - b_4$
$12\beta_5 =$	$-p$	r	$-q$	$-q$	r	$-p$	$2b'_5 = \beta_5 + b_5, \quad 2b'_7 = \beta_5 - b_5$
$12\beta_6 =$	-1	1	-1	1	-1	1	$2b'_6 = \beta_6$

b_k from the analysis problem with 12 points

4. *Trigonometric polynomial.*

$$\pi_{24}(\varphi) = a'_0 + 2(a'_1 \cos \varphi + \ldots + a'_{11} \cos 11\varphi) + a'_{12} \cos 12\varphi$$
$$- 2(b'_1 \sin \varphi + \ldots + b'_{11} \sin 11\varphi)$$

Table 6. Harmonic Synthesis, 12 *l*-Values

1. *Given trigonometric polynomial*

$$y = \pi_{12}(\varphi) = a_0 + 2(a_1 \cos \varphi + \ldots + a_5 \cos 5\varphi) + a_6 \cos 6\varphi$$
$$- 2(b_1 \sin \varphi + \ldots + b_5 \sin 5\varphi)$$

2. *Matrix for the generation of the sums of the I-values.*

	$s_0 =$	$\tfrac{1}{2}s_1 =$	$\tfrac{1}{2}s_2 =$	$\tfrac{1}{2}s_3 =$	$\tfrac{1}{2}s_4 =$	$\tfrac{1}{2}s_5 =$	$s_6 =$
a_0	1	1	1	1	1	1	1
$2a_1$	1	u	0.5		−0.5	$-u$	−1
$2a_2$	1	0.5	−0.5	−1	−0.5	0.5	1
$2a_3$	1		−1		1		−1
$2a_4$	1	−0.5	−0.5	1	−0.5	−0.5	1
$2a_5$	1	$-u$	0.5		−0.5	u	−1
a_6	1	−1	1	−1	1	−1	1

$$u = \cos 30° = 0.8660 \ldots$$

For the evaluation it is convenient to form first the half-sums and half-differences of symmetrically located quantities; for example, $\tfrac{1}{2}s_0 \pm \tfrac{1}{2}s_6$, $\tfrac{1}{4}s_1 \pm \tfrac{1}{4}s_5$. This holds for all following synthesis problems.

3. *Matrix for the generation of the differences of the I-values.*

	$\tfrac{1}{2}d_1 =$	$\tfrac{1}{2}d_2 =$	$\tfrac{1}{2}d_3 =$	$\tfrac{1}{2}d_4 =$	$\tfrac{1}{2}d_5 =$
$2b_1$	−0.5	$-u$	−1	$-u$	−0.5
$2b_2$	$-u$	$-u$		u	u
$2b_3$	−1		1		−1
$2b_4$	$-u$	u		$-u$	u
$2b_5$	−0.5	u	−1	u	−0.5

4. *Generation of the I-values.*

s_0	$\frac{1}{2}s_1$	$\frac{1}{2}s_2$	$\frac{1}{2}s_3$	$\frac{1}{2}s_4$	$\frac{1}{2}s_5$	s_6
	$\frac{1}{2}d_1$	$\frac{1}{2}d_2$	$\frac{1}{2}d_3$	$\frac{1}{2}d_4$	$\frac{1}{2}d_5$	
$y(0°)$	$y(30°)$	$y(60°)$	$y(90°)$	$y(120°)$	$y(150°)$	$y(180°)$
	$y(330°)$	$y(300°)$	$y(270°)$	$y(240°)$	$y(210°)$	

For example, $y(30°) = \frac{1}{2}s_1 + \frac{1}{2}d_1$, $y(330°) = \frac{1}{2}s_1 - \frac{1}{2}d_1$.

Table 7. Harmonic Synthesis, 24 *I*-Values

1. *Given trigonometric polynomial*

$$y = \pi_{24}(\varphi) = a_0' + 2(a_1' \cos\varphi + \ldots + a_{11}' \cos 11\varphi) + a_{12}' \cos 12\varphi$$

$$- 2(b_1' \sin\varphi + \ldots + b_{11}' \sin 11\varphi)$$

a_0'	a_1'	a_2'	a_3'	a_4'	a_5'	a_6'	b_1'	b_2'	b_3'	b_4'	b_5'	b_6'
a_{12}'	a_{11}'	a_{10}'	a_9'	a_8'	a_7'	a_6'	b_{11}'	b_{10}'	b_9'	b_8'	b_7'	b_6'
a_0	a_1	a_2	a_3	a_4	a_5	a_6	β_1	β_2	β_3	β_4	β_5	β_6
α_0	α_1	α_2	α_3	α_4	α_5		b_1	b_2	b_3	b_4	b_5	

For example, $a_1 = a_1' + a_{11}'$, $\alpha_1 = a_1' + a_{11}'$, $a_6 = 2a_6'$.
     For those values of φ which are multiples of 30°, the corresponding y are obtained from the a_k, b_k by means of a synthesis as in table 6.

2. *Generation of the sums of the remaining I-values.*

	$\frac{1}{2}\sigma_1 =$	$\frac{1}{2}\sigma_2 =$	$\frac{1}{2}\sigma_3 =$	$\frac{1}{2}\sigma_4 =$	$\frac{1}{2}\sigma_5 =$	$\frac{1}{2}\sigma_6 =$	
α_0	1	1	1	1	1	1	
$2\alpha_1$	p	r	q	$-q$	$-r$	$-p$	$p = \cos 15°$
$2\alpha_2$	u		$-u$	$-u$		u	$q = \sin 15°$
$2\alpha_3$	r	$-r$	$-r$	r	r	$-r$	$r = \sin 45°$
$2\alpha_4$	0.5	-1	0.5	0.5	-1	0.5	$u = \cos 30°$
$2\alpha_5$	q	$-r$	p	$-p$	r	$-q$	

3. *Generation of the differences of the remaining I-values.*

	$\frac{1}{2}\delta_1 =$	$\frac{1}{2}\delta_2 =$	$\frac{1}{2}\delta_3 =$	$\frac{1}{2}\delta_4 =$	$\frac{1}{2}\delta_5 =$	$\frac{1}{2}\delta_6 =$
$2\beta_1$	$-q$	$-r$	$-p$	$-p$	$-r$	$-q$
$2\beta_2$	-0.5	-1	-0.5	0.5	1	0.5
$2\beta_3$	$-r$	$-r$	r	r	$-r$	$-r$
$2\beta_4$	$-u$		u	$-u$		u
$2\beta_5$	$-p$	r	$-q$	$-q$	r	$-p$
β_6	-1	1	-1	1	-1	1

4. *Function values.*

$\frac{1}{2}\sigma_1$ $\frac{1}{2}\delta_1$	$\frac{1}{2}\sigma_2$ $\frac{1}{2}\delta_2$	$\frac{1}{2}\sigma_3$ $\frac{1}{2}\delta_3$	$\frac{1}{2}\sigma_4$ $\frac{1}{2}\delta_4$	$\frac{1}{2}\sigma_5$ $\frac{1}{2}\delta_5$	$\frac{1}{2}\sigma_6$ $\frac{1}{2}\delta_6$
$y(15°)$	$y(45°)$	$y(75°)$	$y(105°)$	$y(135°)$	$y(165°)$
$y(345°)$	$y(315°)$	$y(285°)$	$y(255°)$	$y(225°)$	$y(195°)$

Table 8. Nomogram for the Cubic Interpolation and for the Construction of the Quadratic Improvement

(Theory 7.11, formula (432) and 7.12)

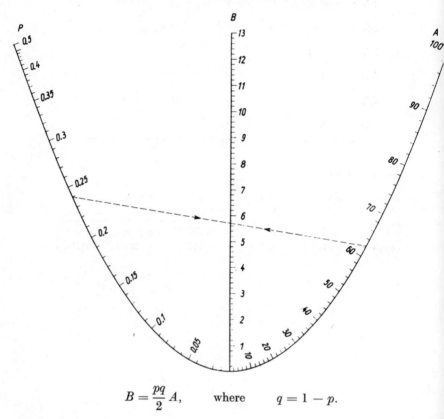

$$B = \frac{pq}{2}\, A, \qquad \text{where} \qquad q = 1 - p.$$

Example for using the monogram:

$$p = 0.24, \qquad A = 62.5, \qquad B = 5.7.$$

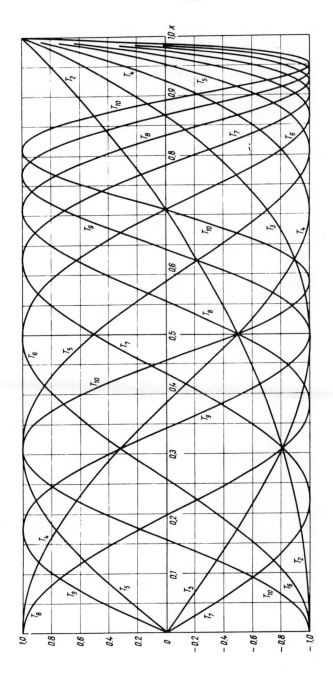

Table 9. Chebyshev Polynomials T_2 Through T_{10} in the Interval from 0 to 1

Literature-References

1. R. Zurmühl, "Praktische Mathematik für Ingenieure und Physiker," 4th ed. Springer, Berlin, 1963 (in German).
2. R. Zurmühl, "Matrizen," 2nd ed. Springer, Berlin, 1958 (in German); 3rd ed. Springer, Berlin, 1961 (in German).
3. Fr. A. Willers, "Methoden der praktischen Analysis," 2nd ed. de Gruyter, Berlin, 1950 (in German). English translation of 1st edition by R. T. Beyer: "Practical Analysis." Dover, New York, 1948.
4. S. I. Gass, "Linear Programming." McGraw-Hill, New York, 1958.
5. S. Vajda, "Readings in Linear Programming." Methuen London, 1958. German translation by P. Künzi: "Lineare Programmierung, Beispiele." Zürich, 1960.
6. J. D. Williams, "The Compleat Strategyst." McGraw-Hill, New York, 1954.
7. F. Lösch, "Siebenstellige Tafeln der elementaren transzendenten Funktionen." Springer, Berlin, 1954 (in German).
8. E. Jahnke, F. Emde, and F. Lösch, "Tables of Higher Functions," 6th ed. McGraw-Hill, New York, 1960 ["Tafeln höherer Funktionen," 6th ed. Teubner, Stuttgart, 1960 (in German)].
9. P. Montel, "Leçons sur les récurrences et leurs applications." Gauthier-Villars, Paris, 1957.
10. H. P. Künzi and W. Krelle, "Nichtlineare Programmierung." Springer, Berlin, 1961 (in German).
11. K. J. Arrow, L. Hurwicz, and H. Uzawa, "Studies in Linear and Non-linear Programming," Stanford Mathematical Studies in the Social Sciences, Vol. II. Stanford Univ. Press, Stanford, California, 1958.
12. A. Ostrowski, "Vorlesungen über Differential- und Integralrechnung," Vol. 1, 2nd ed., 1960; Vol. 2, 2nd ed., 1960; Vol. 3, 1st ed., 1954. Birkhäuser, Basel (in German).
13. H. Rutishauser, "Der Quotienten-Differenzen-Algorithmus." Mitt. Inst. angew. Math. Zürich, No. 7 (1957) (in German).
14. A. S. Householder, "Principles of Numerical Analysis." McGraw-Hill, New York, 1953.
15. F. L. Bauer, Das Verfahren der Treppeniteration und verwandte Verfahren zur Lösung algebraischer Eigenwertprobleme. *Z. angew. Math. u. Phys.* 8 (1957), p. 214–234 (in German).
16. M. Engeli, Th. Ginsburg, H. Rutishauser, and E. Stiefel, "Refined iterative methods for the computation of the solution and the eigenvalues of self-adjoint boundary value problems". Mitt. Inst. angew. Math. Zürich, No. 8 (1959).
17. C. Lanczos, An iteration method for the solution of the eigenvalue problem of linear

differential and integral operators. *In* "Proceedings of a second symposium on large scale digital computing machinery." Ann. Comput. Laboratory, Harvard University, No. 26, Cambridge, Massachusetts, 1951.

18. E. Bodewig, "Matrix Calculus," 2nd ed. North-Holland Publ. Co., Amsterdam, 1959.

19. L. Collatz, "Eigenwertaufgaben mit technischen Anwendungen." Akad. Verlagsges., Leipzig, 1949 (in German).

20. P. Henrici, "Discrete Variable Methods in Ordinary Differential Equations." Wiley, New, York, 1962.

21. L. Collatz, "Numerische Behandlung von Differentialgleichungen," 2nd ed. Springer, Berlin, 1955 (in German); 3rd ed., 1960 (in English) (see additional literature references).

22. H. Rutishauser, Bemerkungen zur numerischen Integration gewöhnlicher Differentialgleichungen n-ter Ordnung. *Numerische Math.* **2** (1960), 263–279 (in German).

23. L. Bieberbach, "Theorie der gewöhnlichen Differentialgleichungen auf funktionentheoretischer Grundlage dargestellt." Springer, Berlin, 1953 (in German).

24. G. Hellwig, "Partielle Differentialgleichungen." Teubner, Stuttgart, 1960 (in German).

25. R. Sauer, Anfangswertprobleme bei partiellen Differentialgleichungen," 2nd ed. Springer, Berlin, 1958 (in German).

26. E. Stiefel, Über discrete und lineare Tschebyscheff-Approximationen. *Numerische Math.* **1** (1959), 1–28 (in German).

27. "Tables of Lagrangian Interpolation Coefficients." Natl. Bur. Standards (U.S.) Columbia Univ. Press Series, No. 4, 1948.

28. L. Collatz, "Differentialgleichungen für Ingenieure," 2nd ed. Teubner, Stuttgart, 1960.

29. " Tables of Chebyschev Polynomials $S_n(x)$ and $C_n(x)$." Natl. Bur. Standards (U.S.) Appl. Math. Series, No. 9, 1952.

30. G. Szego, "Orthogonal Polynomials," revised edition. Amer. Math. Soc., Colloquium Publications, Vol. 23, New York, 1959.

31. J. G. Malkin, "Theorie der Stabilität einer Bewegung." Original Russian edition, Moscow-Leningrad, 1952. German translation by W. Hahn and R. Reissig: Oldenbourg, Munich, 1959.

32. H. J. Maehly, Methods for fitting rational approximations, Part I. Telescoping procedures for continued fractions. *J. Assoc. Computing Machinery* **7**, No. 2 (1960), 150–162.

33. G. E. Forsythe and W. R. Wasow, "Finite Difference Methods for Partial Differential Equations." Wiley, New York, 1960.

34. S. Dano, "Linear Programming in Industry, Theory and Applications." Springer, Vienna, 1960.

35. V. S. Rjabenki and A. F. Filippow, "Über die Stabilität von Differenzengleichungen," Original Russian edition, Moscow, 1956. German translation: Deut. Verlag. Wiss., Berlin, 1960.

36. E. Stiefel, Altes und Neues über numerische Quadratur. *Z. angew. Math. u. Mech.* (*ZAMM*) **41**, 10/11 (1961) (in German); F. L. Bauer, H. Rutishauser, and E. Stiefel, New aspects in numerical quadrature. *In* "Interactions between Mathematical Research and High Speed Computing," Proceedings of a Symposium in Applied Mathematics, Vol. 5, Am. Math. Soc., Providence, Rhode Island (in press).

Additional Literature-References in English

General Texts

S. H. Crandall, "Engineering Analysis." McGraw-Hill, New York, 1956.

R. W. Hamming, "Numerical Methods for Scientists and Engineers." McGraw-Hill, New York, 1962.

P. Henrici, "Lecture Notes on Elementary Numerical Analysis." Prepared for use at the Summer Institute for Numerical Analysis, Univ. of California at Los Angeles. Wiley, New York, 1962.

Z. Kopal, "Numerical Analysis." Wiley, New York, 1955.

K. S. Kunz, "Numerical Analysis." McGraw-Hill, New York, 1957.

W. E. Milne, "Numerical Calculus." Princeton Univ. Press, Princeton, New Jersey, 1950.

A. Ralston and H. S. Wilf (eds.), "Mathematical Methods for Digital Computers." Wiley, New York, 1960.

J. R. Scarborough, "Numerical Mathematical Analysis," 5th ed. Johns Hopkins Press, Baltimore, Maryland, 1962.

J. Todd (ed.), "Survey of Numerical Analysis." McGraw-Hill, New York, 1962.

Linear Algebra

R. Bellman, "Introduction to Matrix Analysis." McGraw-Hill, New York, 1960.

R. A. Frazer, W. J. Duncan, and A. R. Collar, "Elementary Matrices." Cambridge Univ. Press, London, 1938.

V. N. Faddeeva, "Computational Methods in Linear Algebra" (C. D. Benster, translator). Dover, New York, 1959.

M. Marcus, "Basic Theorems in Matrix Theory." Natl. Bur. Standards (U.S.) Appl. Math. Series, No. 57, 1960.

Linear Programming

R. Bellman, "Dynamic Programming." Princeton Univ. Press, Princeton, New Jersey, 1957.

G. Hadley, "Linear Programming." Addison-Wesley, Reading, Massachusetts, 1962.

S. Karlin, "Mathematical Methods and Theory of Games, Programming and Economics," 2 volumes. Addison-Wesley, Reading, Massachusetts, 1959.

J. C. C. McKinsey, "Introduction to the Theory of Games." McGraw-Hill, New York, 1952.

S. Vajda, "An Introduction to Linear Programming and the Theory of Games." Wiley, New York, 1960.

S. Vajda, "Mathematical Programming." Addison-Wesley, Reading, Massachusetts, 1961.

Non-Linear Algebra

A. M. Ostrowski, "Solution of Equations and Systems of Equations."   Academic Press, New York, 1960.

Differential Equations

A. A. Bennett, W. E. Milne, and H. Bateman, "Numerical Integration of Differential Equations."   Dover, New York, 1956.

L. Collatz, "The Numerical Treatment of Differential Equations," 3rd ed.   Springer, Berlin, 1960.

W. E. Milne, "Numerical Solution of Differential Equations."   Wiley, New York, 1953.

I. G. Petrovskii, "Lectures on Partial Differential Equations."   English translation: Interscience, New York, 1954.

R. S. Varga, "Matric Iterative Analysis."   McGraw-Hill, New York, 1962.

Approximation, Interpolation

R. E. Langer (ed.), "On Numerical Approximation," Proceedings of a Symposium conducted by the Math. Research Center, United States Army, at the University of Wisconsin, 1958.   Univ. of Wisconsin Press, Madison, Wisconsin, 1959.

J. Todd, "Constructive Function Theory."   Birkhäuser, Basel (in press).

Tables

C. W. Clenshaw, "Chebyshev Series for Mathematical Functions."   Math. Tab. Nat. Phys. Lab 5, H. M. Stationery Office, London, 1963.

L. J. Comrie, "Chamber's Six-Figure Mathematical Tables," two volumes.   Chambers, Edinburgh, Scotland, 1948; Van Nostrand, Princeton, New Jersey, 1949.

"Interpolation and Allied Tables."   Nautical Almanac Office, H. M. Stationery Office, London, 1956.

E. Jahnke and F. Emde, "Tables of Functions with Formulae and Curves," 4th ed.   Dover, New York, 1945 (text in both English and German).

Author Index

Subject Index

283